I0787951

Silicon Compatible Emerging Materials, Processes, and Technologies for Advanced CMOS and Post-CMOS Applications 9

Editors:

F. Roozeboom

P. J. Timans

K. Kakushima

E. P. Gusev

Z. Karim

D. Misra

Y. S. Obeng

S. De Gendt

H. Jagannathan

Sponsoring Divisions:

 Electronics and Photonics

 Dielectric Science and Technology

Published by
The Electrochemical Society
65 South Main Street, Building D
Pennington, NJ 08534-2839, USA

tel 609 737 1902
fax 609 737 2743
www.electrochem.org

ecstransactions ™

Vol. 89, No. 3

Copyright 2019 by The Electrochemical Society.
All rights reserved.

This book has been registered with Copyright Clearance Center.
For further information, please contact the Copyright Clearance Center,
Salem, Massachusetts.

Published by:

The Electrochemical Society
65 South Main Street
Pennington, New Jersey 08534-2839, USA

Telephone 609.737.1902
Fax 609.737.2743
e-mail: ecs@electrochem.org
Web: www.electrochem.org

ISSN 1938-6737 (online)
ISSN 1938-5862 (print)
ISSN 2151-2051 (cd-rom)

ISBN 978-1-62332-565-7 (CD-ROM)
ISBN 978-1-62332-566-4 (USB)
ISBN 978-1-60768-868-6 (PDF)

Printed in the United States of America.

PREFACE

From May 26 to 30, 2019, the international symposium on "Silicon Compatible Materials, Processes, and Technologies for Advanced Integrated Circuits and Emerging Applications 9" was held as part of the 235th Meeting of The Electrochemical Society. The venue was Dallas, Texas, known as the hometown of Texas Instruments, where the legendary Nobel Prize winner Jack Kilby invented the first integrated circuit over 60 years ago.

Befitting to the venue, this year's annual symposium focused on advanced materials, devices, and technologies for enabling the continued scaling of CMOS integrated circuits. The symposium featured two major awards from the Society along with talks from their respective recipients. The first award talk, on silicon-based photonic integrated circuits, was given by David Lockwood (National Research Council, Canada), who received the Gordon E. Moore Medal for Outstanding Achievement in Solid State Science and Technology. The second awardee was Sean King (Intel), who received the Thomas D. Callinan Award from the ECS Dielectric Science and Technology Division. His talk featured advancements in the field of dielectric science with his perspectives and forecast of the field.

Aligned to the transforming technological landscape, this symposium continues to evolve. This year, the symposium featured several papers focusing on novel nonvolatile memory/switching elements, materials and devices for neuromorphic computing, and machine/artificial intelligence. This issue of *ECS Transactions* contains a selection of papers on new materials, processes, and devices in semiconductor technology, photonics, and nonvolatile memory presented at the symposium. The organizers would like to express their gratitude to all speakers, invited and contributing, for their interest in this symposium, and for submitting high-quality abstracts and preparing their manuscripts. Finally, the success of the symposium is greatly and positively influenced by the financial support given by the sponsoring ECS divisions—Electronics and Photonics (lead sponsor) and Dielectric Science and Technology—as well as our industrial sponsor, IBM. Their loyal support and sponsorship are highly appreciated.

F. Roozeboom, Eindhoven University of Technology and TNO Holst Centre, Eindhoven, The Netherlands

P. J. Timans, Thermal Process Solutions Limited, Cambridge, United Kingdom

K. Kakushima, Tokyo Institute of Technology, Yokohama, Japan

E. P. Gusev, Qualcomm Technologies, Inc., Santa Clara, California, USA

Z. Karim, Yield Engineering Systems Inc., Livermore, California, USA

D. Misra, New Jersey Institute of Technology, Newark, New Jersey, USA

Y. Obeng, National Institute of Standards and Technology, Gaithersburg, Maryland, USA

S. DeGendt, IMEC and Catholic University Leuven, Leuven, Belgium

H. Jagannathan, IBM Research, Albany, New York, USA

May 2019

***ECS Transactions*, Volume 89, Issue 3**
Silicon Compatible Emerging Materials, Processes, and Technologies
for Advanced CMOS and Post-CMOS Applications 9

Table of Contents

Chapter 1
Gordon E. Moore Medal for Outstanding Achievement in Solid State Science and Technology Award Address

Chapter 2
Materials and Devices for Novel Non-volatile Memory Elements and Neuromorphic Computing

Chapter 3
Poster Session

Facts about ECS

The Electrochemical Society (ECS) is an international, nonprofit, scientific, educational organization advancing the theory and practice of electrochemistry and solid state science and technology, and allied subjects. The Society was founded in Philadelphia in 1902 and incorporated in 1930. There are currently over 8,400 members from around the globe representing 13 technical division and 23 geographical sections and a growing student membership program with almost 70 student chapters. The Society is also supported by more than 800 corporations, government agencies and academic institutions through institutional membership, corporate programs and subscriptions.

The technical activities of the Society are carried on by divisions. Sections of the Society host symposia, programs and events focused on their respective geographic regions. Major international meetings of the Society are held in the spring and fall of each year. At these meetings, the divisions and partnered organizations hold general sessions and sponsor symposia on specialized subjects.

The Society has an active publication program that includes the following:

Journal of The Electrochemical Society — (JES) is the flagship journal of The Electrochemical Society and the oldest peer-reviewed journal in its field. Since its founding in 1902, JES has evolved into one of the most highly cited and prestigious journals in electrochemistry and materials science with a cited half-life of greater than 10 years.

ECS Journal of Solid State Science and Technology — (JSS) is one of The Electrochemical Society's newest peer-reviewed journals. Launched in 2012, JSS covers fundamental and applied areas of solid state science and technology, including experimental and theoretical aspects of the chemistry and physics of materials and devices.

ECS Transactions — (ECST) is an online database of technical papers—a high-quality venue for authors and an excellent resource for researchers. ECST offers the full-text content of proceedings from ECS meetings and ECS sponsored conferences.

The Electrochemical Society Interface — *Interface* is a quarterly news magazine that provides a forum for the lively exchange of ideas within the scientific community. Published online with free access in the ECS Digital Library, issues engage readers with a featured scientific research section, and summaries of research breakthroughs, and industry and Society news.

ECS Meeting Abstracts — Published 2-4 times a year, this publication contains extended abstracts of technical papers presented at ECS biannual meetings and ECS-sponsored meetings and offers a first look into the current research in the field.

ECS Monograph Series — ECS monographs provide authoritative, detailed accounts of specific topics in electrochemistry and solid state science and technology. These titles are sponsored by ECS and published in cooperation with noted publishers such as John A. Wiley & Sons.

For more information on these and other Society activities, visit the ECS website:

www.electrochem.org

Chapter 1

Gordon E. Moore Medal for Outstanding Achievement in Solid State Science and Technology Award Address

ECS Transactions, 89 (3) 3-35 (2019)
10.1149/08903.0003ecst ©The Electrochemical Society

Towards Silicon-Based Photonic Integrated Circuits:
The Quest for Compatible Light Sources

David J. Lockwood

Measurement Science and Standards, National Research Council Canada,
Ottawa, Ontario K1A 0R6, Canada

Light emission from silicon and germanium nanostructures has been of great interest for some time now owing to the need for silicon-based light sources for applications in silicon optoelectronics and photonics. Both silicon and germanium possess indirect band gaps, which makes them very inefficient light emitters. Band gap engineering employing quantum wells, wires, or dots has been proposed as one way to overcome this limitation and, in the past, Si/Ge, Si/SiO$_2$, or Si/SiGe-alloy thin-multilayer nanostructures grown on Si have been produced on this principle, and although light emission with a greatly improved efficiency has been obtained at low temperatures the emission at room temperature is still very weak, because of exciton dissociation. Recent advances in band gap engineering technology have resulted in the development of different systems incorporating one-, two-, and three-dimensionally confined Si-Ge nanostructures that produce bright light in the visible to near infrared. Here, the novel optical properties of various such nanostructures are reviewed.

Introduction

Light emission from Si and Ge nanostructures (NSs) has been of great interest in recent years (1–4) owing to the need for silicon-based light sources for applications in silicon optoelectronics and photonics (5,6). Optical interconnects are required these days for on-chip technology as an alternative to metal wires, because of data transmission bottlenecks introduced by their unavoidable delay times, significant signal degradation, problems with power dissipation, and electromagnetic interference. Two major avenues toward optical interconnects on a chip include a hybrid approach with III-V densely packaged optoelectronic components and the all-group-IV approach (mainly Si, Ge and SiGe), where all the major components, e.g., light emitters, modulators, waveguides and photodetectors, are monolithically integrated into the CMOS environment (5,6). Both Si and Ge possess indirect band gaps, which makes them very inefficient light emitters (1). Band gap engineering employing quantum wells, quantum wires or quantum dots has been proposed as one way to overcome this limitation and Si/Ge or Si/SiGe-alloy thin multilayer quantum well structures grown on Si have been produced on this principle, and although light emission with greatly improved efficiency has been obtained at low temperatures the emission at room temperature is still very weak, because of exciton dissociation (1,5,7). Recently, through employing novel band-gap engineering strategies, several different entirely new bright light-emitting Si/Ge nanostructures including one possessing a direct gap have been prepared. The latter structure is based on constructing a new super unit cell comprised of multiple planar epitaxial layers of Si and Ge grown on (001) $Si_{0.4}Ge_{0.6}$ (8). Others are based on silicon-germanium layers grown epitaxially on silicon in such a way as to form multiple layer three-dimensional nanostructures (quantum dots) (9,10). A simple and efficient electrochemical process that combines galvanic reaction and focused-ion-beam lithography to selectively synthesize gold nanoparticles has been developed that can be used to grow ordered SiGe nanowire arrays with a predefined diameter (200 nm) and position (11,12). Here, the light emitting properties of these and other similar Si/Ge nanostructures that have been found to luminesce efficiently at wavelengths in the important spectral range of 1.1–1.6 μm are compared for their possible use in optoelectronics and photonics. Emphasis is placed on the work carried out at the National Research Council of Canada over the last three decades.

Optical Evidence for Quantum Confinement Effects in Porous Silicon

Porous silicon (p-Si) was discovered over 60 years ago by Uhlir at Bell Telephone Laboratories, USA (13,14). The porous material is created by electrochemical dissolution in HF-based electrolytes. Hydrofluoric acid, on its own, etches single-crystal Si extremely slowly, at a rate of only nanometers per hour. However, passing an electric current between the acid electrolyte and the Si sample speeds up the process considerably, leaving an array of deep narrow pores that generally run perpendicular to the Si surface. Pores measuring only nanometers across, but micrometers deep, have been achieved under specific etching conditions (15,16). The evenly-spaced pore structure was explained in terms of available paths for the etching current, with each pore surrounded by an insulating layer of material depleted of electrons. After many years of basic research on the chemical and physical properties of p-Si, the formation mechanism of the network of pores is now largely understood and the resulting porous microstructure depends critically not only on the HF acid concentration and anodization parameters (current density or voltage) but also on the type and doping level of the Si crystal (17,18). The Si skeleton left behind has been shown to retain the substrate crystallinity, but with a small lattice expansion compared to bulk Si (19). In the mid-1970s, p-Si was already being considered for use as a dielectric in the development of silicon-on-insulator (SOI) technology for very-large-scale integrated circuits (17,20,21). It is only comparatively recently that the optical characteristics of p-Si have been explored in detail. In 1984, Pickering *et al.* (22) reported on low temperature luminescence in p-Si anodized under a range of conditions and noted a band at higher energy (above the bulk-Si band gap) that shifted up in energy with decreasing p-Si density, but they did not investigate the emission properties at room temperature. The complexities that have arisen in the interpretation of the photoluminescence (PL) were already apparent in this fundamental work.

In July 1989, Canham conceived the idea of fabricating Si quantum wires in p-Si by reverting to the much slower chemical HF etch after the electrochemical etch (23). In this way he proposed to join up the pores leaving behind an irregular array of undulating free standing pillars of crystalline Si (c-Si) only nanometers wide. In 1990, Canham (24) remarkably observed intense visible PL at room temperature in p-Si that had been etched under carefully controlled conditions. Visible luminescence ranging from green to red in color was soon reported by Canham *et al.* for other p-Si samples and ascribed to quantum size effects in wires of width ~ 3 nm (24,25). Independently, Lehmann and Gösele (26) reported on the optical absorption properties of p-Si. They observed a shift in the bulk Si absorption edge to values as high as 1.76 eV that they also attributed to quantum wire formation. Tremendous activity on research into the physical and associated chemical characteristics of p-Si ensued from these early reports with, unfortunately, some duplication of effort, and a great diversity in their luminescence properties was observed (27). At the National Research Council we quickly joined in this momentous work and from a detailed investigation of the optical properties of p-Si we concluded that there is strong optical absorption evidence for quantum confinement effects in c-Si NSs within p-Si and that the red-orange PL also exhibits a weaker quantum confinement effect (28–33).

Transmission electron microcopy and Raman scattering spectroscopy has shown that in our p-Si samples the porous material is in the form of mainly sphere-like nanocrystals, and thus we refer to the nanostructures as spherites (28). A typical optical transmission curve obtained by at room temperature from a p-Si membrane sample is shown in Fig. 1 together with the corresponding PL spectrum (30). It is very noticeable that the PL peak energy is at a lower energy than the optical absorption edge. This behavior is even more evident when the band gap energy obtained from the optical absorption spectrum is compared for a number of samples of different average diameter, as shown in Fig. 2 (30).

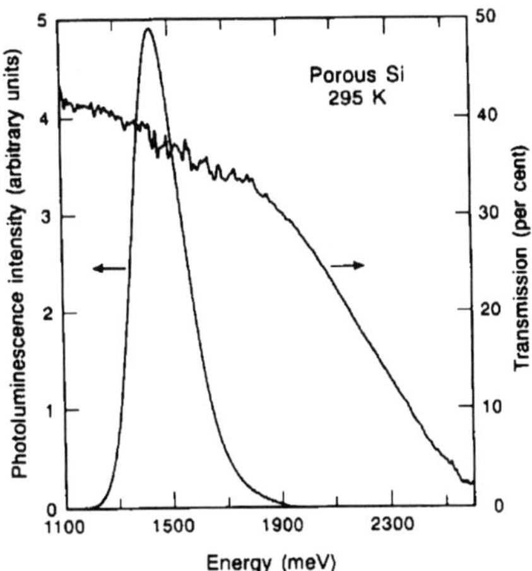

Figure 1. Room temperature transmittance and PL spectra for a p-Si membrane comprised of 3.1 nm spherites.

The PL data in Fig. 2 is in remarkably good agreement with the predictions of simple effective mass theory for quantum dots (28) and this justifies further interpretation within quantum confinement models. A more sophisticated calculation within a linear combination of atomic orbitals (LCAO) framework for c-Si spheres with hydrogen passivated surfaces (34) has predicted a $d^{-1.39}$ dependence of the optical gap on the sphere diameter, d. Other calculations for dots (35,36) in a variational envelope function approach produced similar results. The calculated energy gap behavior (34,35) has been shown (27) to be in excellent agreement with optical absorption data for p-Si spherites (29,30,37) and c-Si spheres (38). However, the calculation by Proot et al. (25), which is such a good fit to our absorption data, lies generally above the PL data, as shown in Fig. 2. A similar LCAO calculation has been undertaken for c-Si wires (39) and the predicted energy gap dependence on the wire diameter, which is similar to that obtained from several other theories (27), is also plotted in Fig. 2. The wire theory curve is in closer agreement with the p-Si PL data, as could be expected. However, to permit this more satisfactory comparison with quantum wire theory the extreme and unlikely case of *uniform* wires has to be taken, and thus the energy difference between the absorption and emission data remains unexplained.

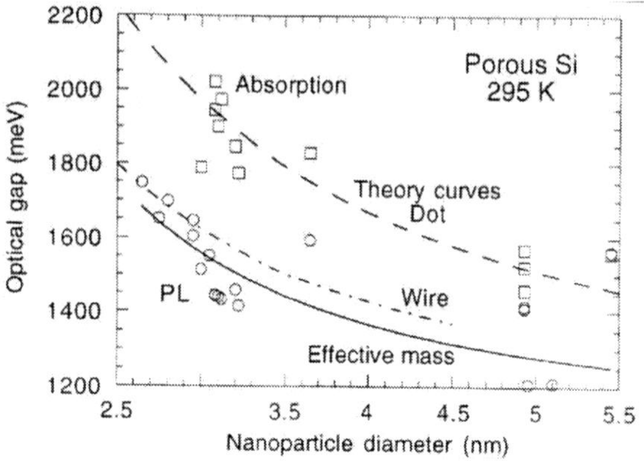

Figure 2. Spherite diameter dependence of the PL peak energy and optical absorption in p-Si samples at room temperature. The solid line is the effective mass model prediction for the optical energy gap in c-Si spheres, while the broken and short dash-dot lines are the theoretical predictions for quantum dots and wires, respectively.

The difference between the predicted energy gap for spheres, as confirmed in our experiments, and the lower energy of the PL peak has to be attributed to the emission mechanism. In the vibronic model, this difference is associated with phonon-assisted transitions employing c-Si phonons of energy ~65 meV. The difference between dot theory and experiment evident in Fig. 2 is ~100 meV at 5 nm nanoparticle diameter, but it *increases* with decreasing diameter to ~400 meV at 3 nm, and is thus incompatible with a vibronic model. That is because in the vibronic model, multiphonon assisted transitions are required where the average number of phonons involved would have to increase with deceasing nanoparticle diameter, which is not normally the case. For our samples, some additional energy loss must occur in the absorption-emission process prior to emission to account for this large difference at small diameters. As the samples have all been treated in the same way after electrochemical etching, we thought at the time that it was unlikely that the blue shift in the PL with decreasing nanoparticle diameter is due to some change in the surface chemistry, and that this would seem to rule out explanations based on surface recombination models. We considered other models such as the formation of bound excitons and the creation of localized excitons due to trapping by donor-like and acceptor-like states (29). It was not until much later that the true cause of the red PL observed in most p-Si samples was identified (40). It was, in fact, related to a specific surface defect in p-Si created by surface oxidation subsequent to drying the sample – new electronic states appear in the band gap of the smaller quantum dots when a Si=O bond is formed – that produced PL at the observed energy and also exhibited a weak quantum confinement effect. The true intrinsic PL in p-Si is thus only observable when the surface oxidation can be eliminated though effective surface passivation by other methods, such as with Si-H bonds (40) or thermal (non-catalytic) reaction with alkenes and aldehydes (41–46). The latter method leads to high stability (41–46).

Despite this early promise as a room-temperature light emitter, p-Si has not yet been employed as a light emitter in optoelectronics due to the difficulty of integrating it into CMOS production.

Disordered Silicon Quantum Wells

The simplest approach to investigating quantum confinement effects in Si is to grow well-defined thin quantum wells of Si separated by wide band gap barriers. Suitable barrier candidates are SiO_2, CaF_2, and Al_2O_3 (47), and SiO_2 has the additional advantage of being an excellent passivator of Si (48). Although a number of Si/barrier superlattices have been produced in the past (49) none had produced convincing evidence for quantum confinement induced emission until 1995. In 1995, Lu *et al.* (50) reported visible light emission at room temperature from ultrathin-layer Si/SiO_2 superlattices grown by MBE that exhibited a clear quantum confinement shift with Si layer thickness, as shown in Fig. 3. According to effective mass theory and assuming infinite potential barriers, which is a reasonable approximation since wide-gap (9 eV) 1-nm-thick SiO_2 barriers are used, the energy gap E for one-dimensionally confined Si should vary as

$$E = E_g + \frac{\pi \hbar^2}{2d^2}\left(\frac{1}{m_e^*} + \frac{1}{m_h^*}\right), \tag{1}$$

where E_g is the bulk material band gap and m_e^* and m_h^* are the electron and hole effective masses (51). This simple model is a reasonable first approximation to compare with experiment for quantum wells (52). The shift in PL peak energy with Si well thickness d is well represented by Eq. (1), as can be seen in Fig. 3, with $E(eV) = 1.60 + 0.72d^{-2}$. The very thin layers of Si ($1 < d < 3$ nm) have a disordered (d-Si), but nearly crystalline, structure owing to the growth conditions and the huge strain at the Si–SiO_2 interfaces (53). However, by reducing considerably the SiO_2 thickness below 1 nm, epitaxial growth of the Si layers can be achieved via small holes in the SiO_2 film (54). The fitted E_g of 1.60 eV is larger than that expected for c-Si (1.12 eV at 295 K), but is in excellent agreement with that of bulk a-Si (1.5–1.6 eV at 295 K). The indications of direct band-to-band recombination were confirmed by measurements via X-ray techniques of the conduction and valence band shifts with layer thickness (50,55). The fitted confinement parameter of 0.72 eV/nm^2 indicates $m_e^* \approx m_h^* \approx 1$, comparable to the effective masses of c-Si at room temperature (51). The integrated intensity at first rises sharply with decreasing Si thickness until $d \approx 1.5$ nm and then decreases again, which is consistent with quantum well exciton emission (56,57). The PL intensity is enhanced by factors of up to 100 upon annealing and is also selectively enhanced and band-width narrowed by incorporation into an optical microcavity (58). In a higher quality cavity, a very sharp (17 meV full width at half maximum) PL peak has been observed (59).

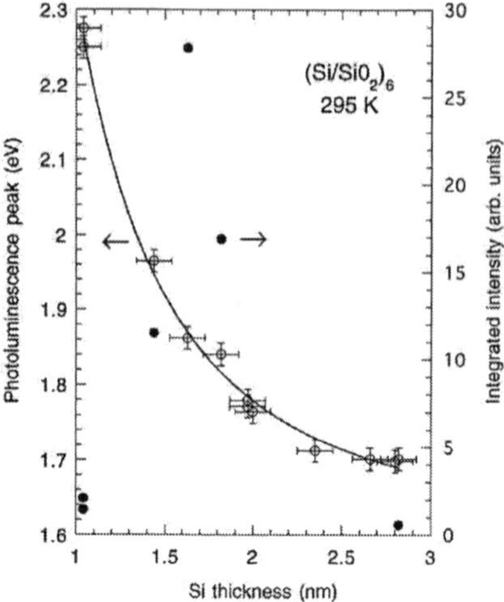

Figure 3. The PL peak energy (open circles) and integrated intensity (full circles) at room temperature in d-Si/a-SiO$_2$ superlattices as a function of Si layer thickness. The solid line is the fit to the peak energy by effective mass theory.

The bright PL obtained from these and other (60) as-grown and annealed Si/SiO$_2$ superlattices (see Fig. 4) offers interesting prospects for the fabrication of a Si-based light emitter that can be tuned from 500 to beyond 800 nm by varying the Si-layer thickness and/or the annealing conditions, all using available vacuum deposition technology and standard Si wafer processing techniques. The next important step is to develop LEDs based on such superlattices. Several prototype devices have been constructed in Si/SiO$_2$ (61–64) and Si/CaF$_2$ (65) and all report visible EL, although there is no strong evidence that the emission originates from confined states in the Si quantum wells. The EL from the Si/SiO$_2$ superlattices is notably stable (61,62).

Figure 4. Room temperature PL from a patterned (d-Si/SiO$_2$)$_{100}$ superlattice, with a periodicity of 2.6 nm, that was post-annealed in air at 1100 °C for 0.5 h. The PL was excited with an Ar laser operating at 457.9 nm and having a power density of ~50 mW/cm^2 at the superlattice surface. The actual size of the logo is ~20×20 mm^2.

Crystalline Silicon Quantum Wells

As discussed above, the d-Si/a-SiO$_2$ system has received the most attention and direct evidence of quantum confinement induced PL has been obtained. However, the as-grown thin Si films (< 3 nm thick) proved very difficult to crystallize by thermal annealing due to strain effects induced by the amorphous SiO$_2$ (a-SiO$_2$) barriers (53,66). In other investigations of single Si quantum-wells formed by the separation by implanted oxygen (SIMOX) process, strong optical recombination of carriers was observed at the c-Si/a-SiO$_2$ interfaces (67–69) together with selective excitation evidence for a weak intrinsic PL (68). Thus, a detailed experimental study of quantum confinement effects in c-Si/a-SiO$_2$ had not been possible for ultrathin silicon layers. Such studies are important for comparison with theory, which is incapable at present of predicting the optical properties of quantum confined d-Si. Then in 2002, Lu and Grozea (70) demonstrated the preparation of single nanometer-thick c-Si/a-SiO$_2$ quantum wells based on oxidizing and etching Canon epitaxial layer transfer (ELTRAN) Si-on-insulator (SOI) wafers. Canon ELTRAN technology uses wafer bonding to fabricate the SOI wafer, which has the benefits of an epitaxial silicon layer that is atomically smooth and free of particles or pits (71). This atomically flat Si/SiO$_2$ interface is essential for electronic applications of SOI wafers and is the key factor in producing silicon quantum wells only a few atomic layers thick. The SOI wafer has a 200 nm thick buried oxide layer surmounted by a transferred epitaxial (100) Si layer about 50 nm thick. To bring the silicon layer thickness down to a few nanometers, furnace dry oxidation was first used to oxidize the Si layer and thereby reduce the c-Si film thickness to 5–10 nm. To produce c-Si films with a thickness of just a few atomic layers, the film was then thinned down atomic-layer-by-atomic-layer by using a combination of room-temperature UV-ozone oxidation and wet chemical etching, according to the method described in Ref. (70). The resulting c-Si film thickness was determined by X-ray photoelectron spectroscopy (70,72).

Representative results obtained from PL measurements at room temperature of such single wells (73–76) are shown in Fig. 5. The overall intensity of the PL was quite weak, but consistent with a single quantum well as opposed to the multiple well samples discussed above. In general, the overall PL intensity increased at first with decreasing layer thickness and then decreased again (see Fig. 5), much as was observed for d-Si wells, (51) while the overall PL line width increased. The temperature dependence of the PL spectrum (77) of one sample is shown in Fig. 6 and these results are quite revealing. They indicate that the PL envelope actually comprises two bands, one at about 1.79 eV at 4 K and the other at around 1.84 eV. (The differences in PL line shape at 300 K, as compared to the spectrum for the same sample given in Fig. 5, arise from the different responsivities of the two sets of PL equipment.) The two bands do not exhibit strong temperature dependences in their integrated intensities, line widths, or positions, but the 1.8 eV band becomes more evident at lower temperatures. Other samples gave similar results with respect to the temperature dependences of the band parameters and also indicated the presence of two bands. The PL line shape could be readily curve-resolved into two bands, as shown in Fig. 5. The presence of these two overlapping bands explains the apparent overall PL intensity and linewidth variation with Si layer thickness.

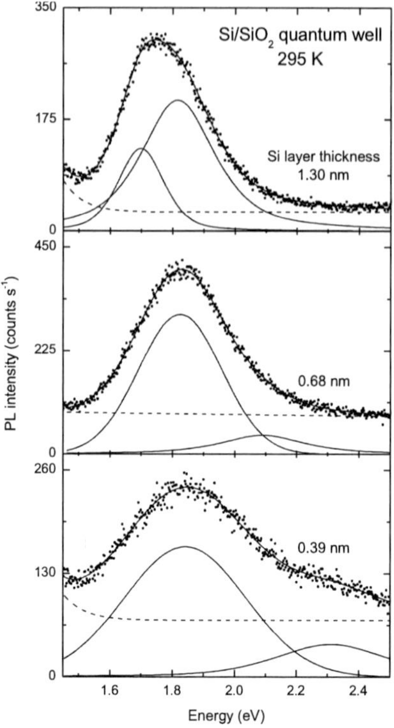

Figure 5. Room temperature PL from single c-Si/a-SiO$_2$ quantum wells of different thickness. The PL line shape has been fitted with the two bands (indicated by the solid line passing through the data points) shown below by the solid lines. The dashed line is the fitted background.

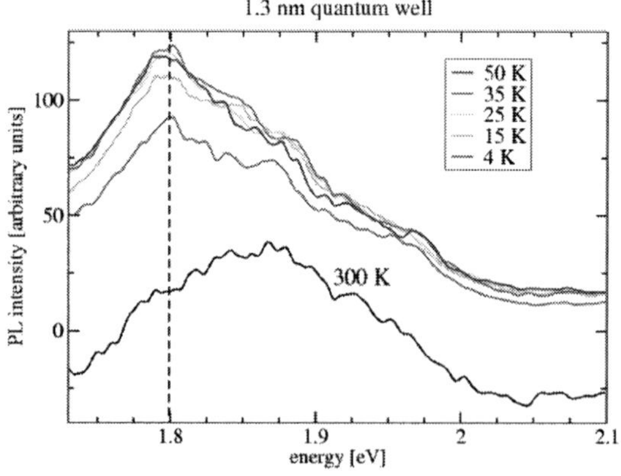

Figure 6. Temperature dependence of the PL from a 1.3 nm thick single c-Si/a-SiO$_2$ quantum well.

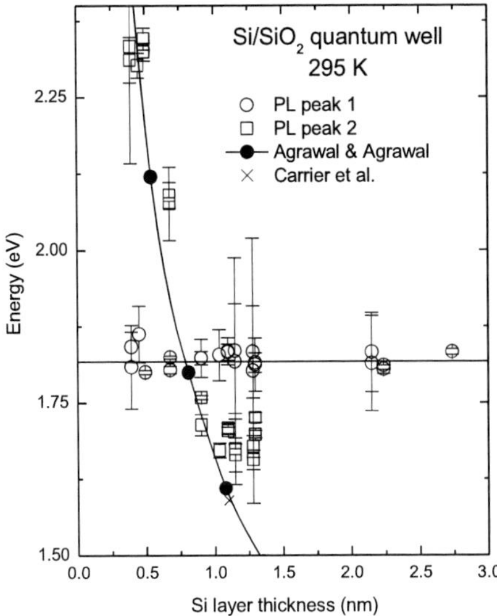

Figure 7. Experimental results (○,□) for the energies of the two PL bands of single c-Si/a-SiO$_2$ quantum wells as a function of well thickness compared with the theory of Agrawal and Agrawal (●) (78) and Carrier *et al.* (×) (79). The solid line is an interpolation of the theoretical data. The error bars represent the standard deviation in the peak energy determined from the fits. The horizontal solid line is a least-squares fit to the data of peak number 1.

The fitted PL band energies as a function of c-Si layer thickness are plotted in Fig. 7. The energy of peak number 1 scarcely varies with Si layer thickness, while peak number 2 exhibits a strong dependence on this thickness. The second band could not be detected for samples with layer thicknesses greater than 1.3 nm, including one sample with a layer thickness of 4.26 nm. This would imply that the band peak energy was lower than the instrumental detection limit of 1.45 eV. The band gap energy variation predicted from theoretical calculations based on self-consistent full potential linear muffin-tin orbital (78) and first-principles projector-augmented wave (79) methods are also shown in Fig. 6. These calculations take into account properly the layered structure (78,79), the SiO$_2$ barriers (79), and the strained c-Si/c-SiO$_2$ interface (79) and differ from the predictions of earlier theoretical models that have relied mostly on a simple H-termination of an isolated Si nanocluster (80,81). In these theories, the band gap energy is that of bulk c-Si (1.1 eV) in the limit of ultrawide quantum wells, which contrasts with that of d-Si (1.6 eV). The experimental results for peak 2 are in remarkable agreement with theory, confirming the origin of this peak as intrinsic quantum-well emission from c-Si and validating the theories. The strong dependence on Si layer thickness evident for thicknesses less than 1.5 nm indicates why peak number 2 was not observed experimentally for larger layer thicknesses (> 2 nm) due to the cut-off in the instrumental response. The higher energy gaps, E_1 and E_2, of c-Si were also investigated by spectroscopic ellipsometry for these samples, but contrary to the fundamental band gap behavior no significant variation with well thickness was observed (82).

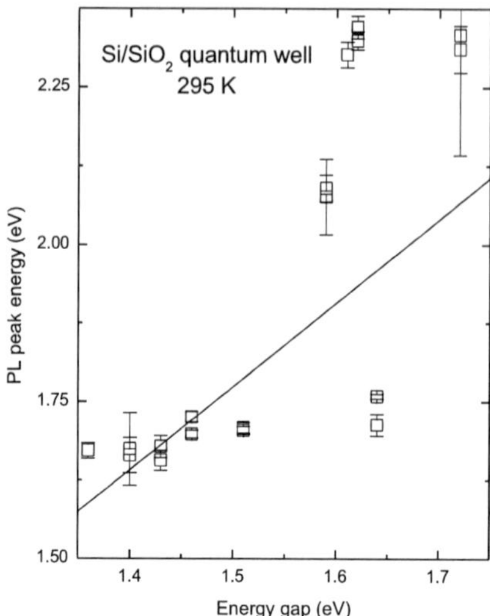

Figure 8. Energy of PL peak number 2 versus the measured energy gap for single c-Si/a-SiO$_2$ quantum wells. The straight line is an error-weighted least-squares fit to the data.

The experimental energy gap for these samples has been determined previously by X-ray techniques, as reported in Ref. 70. The PL energy of peak number 2 increases in accord with the increasing energy gap with decreasing Si layer thickness, as shown in Fig. 8 (75). A least-squares error-weighted linear fit to the data yielded a zero intercept of -0.2 ± 0.8 and a slope of 1.3 ± 0.5 for the relationship between the PL and band-gap energies. This, within error, is the direct proportionality expected for true quantum confinement induced PL (50,83). Thus these experimental results provide the first verification of such theories and vindicate the approaches taken and their associated approximations. These theories can now be applied with confidence to enumerating other properties of the c-Si/SiO$_2$ system and analogous structures.

The data for peak number 1 in Fig. 7 are a good fit to a constant energy of 1.82 eV. Such a behavior negates assignment to light emission within the quantum well. Experimental studies of carrier recombination in oxidized silicon nanostructures have shown similar weak trends with change in Si nanocrystal size (84). In particular Wolkin *et al.* (40) observed a dramatic drop in the confined PL energy for small Si dot diameters (< 2 nm) when the silicon was oxidized and attributed this red shift to recombination involving a trapped electron or exciton. By analogy with these results and others where a PL peak near 1.8 eV is commonly observed (27,84), the PL mechanism in our samples is similarly attributed to trapped state emission at the c-Si/a-SiO$_2$ interface. This mechanism would account for the observed very weak confinement effect on peak number 1.

Subsequent work by Cho *et al.* (69) has confirmed the observation of quantum confinement induced PL from c-Si/a-SiO$_2$ wells produced by high temperature thermal oxidation of ELTRAN SOI wafers, but in this case weak or no interface mediated PL was seen. Most remarkably, stimulated emission from such a single c-Si/a-SiO$_2$ quantum well has been observed by Saito *et al.* (85,86) using electrical pumping in a device constructed entirely with conventional CMOS technology, but so far no lasing action has been reported.

Super Unit Cell

In recent times, through employing novel band gap engineering computations, entirely new Si/Ge (87) and Si (88) structures possessing direct gaps have been proposed. The former structure is based on constructing a new super unit cell comprised of multiple epitaxial layers of Si and Ge grown on (001) $Si_{1-x}Ge_x$ with $x \geq 0.6$ (87) while the latter structure comprises a metastable cubic Si_{20} phase with a quasi-direct band gap of 1.55 eV (88). According to d'Avezac *et al.* (87), a $SiGe_2Si_2Ge_2SiGe_n$ superstructure on $Si_{0.4}Ge_{0.6}$ should have a direct and dipole-allowed gap of 0.863 eV, which is ideally suited for optical fiber data transmission applications.

To investigate the first proposal, two similar samples have been prepared by quite different growth methods: molecular beam epitaxy (MBE) and solid phase epitaxy (SPE). In both samples the $SiGe_2Si_2Ge_2SiGe_{12}$ superstructure was grown on a 5 μm thick relaxed buffer layer of $Si_{0.4}Ge_{0.6}$ on a 750 μm thick (001) Si substrate (8,89).

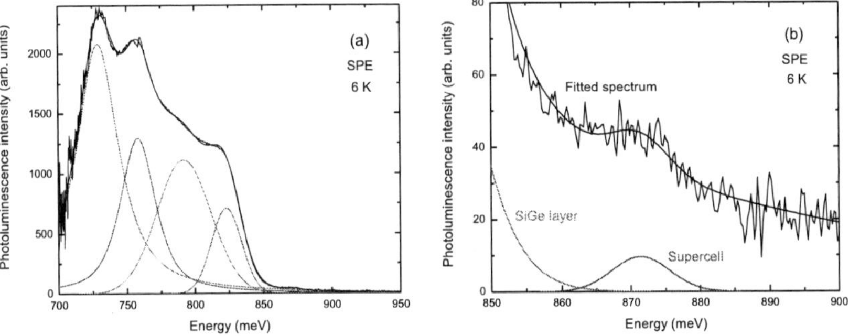

Figure 9. (a) Curve-resolved PL spectrum of the SPE-grown sample at 6 K obtained with 405 nm excitation. (b) Curve-resolving results obtained near the 871-meV PL peak. The solid line shows the fit to the PL data, while component lines are shown underneath.

Characteristic PL results, as obtained for the SPE-grown sample at 6 K, are shown in Fig. 9. Four strong peaks are seen at 728.9, 758.1, 791.8, and 823.1 meV and a much weaker peak at 871.4 meV [Fig. 9(b)]. A fit to the PL results obtained from the same sample at 20 K yielded strong peaks at 724.2, 756.5, 781.6, and 812.6 meV and a weak peak at 869.1 meV. Similar results were obtained from fits to the MBE-grown sample spectra, which were less intense (8,89).

Given the quite close agreement in energy, the weak peak at 0.871 eV is the expected dipole-allowed direct-gap transition predicted to be at 0.863 eV in the superstructure. This peak can be anticipated to be quite weak in PL owing to the fact that this composite layered structure in our samples is only one 'unit cell' thick. Apart from the uncertainty arising from the standard error in the fits (~4 meV), the remaining small difference in energy between theory and experiment could easily be the result of a difference in strain within the layer in the sample compared with the ideal (perfect) modeled structure or arise from assumptions made for parameter values in the model.

Because of their high intensity, the other four peaks must arise from the $Si_{0.4}Ge_{0.6}$ relaxed buffer layer, which is the thickest component of the structure apart from the Si substrate itself that contributes only very weakly to the PL. The energies of the peaks, however, are much lower than that expected for a bulk $Si_{0.4}Ge_{0.6}$ alloy where the indirect energy gap is 971 meV (90). The energy separations and general appearance of these peaks is reminiscent of PL that has been attributed to the presence of various dislocation defects (i.e., the four so-called D-lines) in Si (91–94) and SiGe alloys (91). These PL features are quite sharp in Si (91,92), but are much broader in higher Ge-concentration alloys (95). It is therefore most likely that this PL arises predominately from dislocations formed during the growth of the relaxed alloy buffer layer on the Si substrate.

In summary, experimental evidence has been obtained of the predicted direct-gap optically-allowed transition in a specially engineered supercell comprised of a number of ultrathin layers of Si and Ge. For two quite differently prepared samples, a peak is observed at 0.871 eV, which is very close in energy to the theoretically predicted direct gap of 0.863 eV for this particular structure. Unfortunately, from a room temperature device point of view, the peak is only observable presently at low temperatures (< 25 K).

Three-Dimensional SiGe Nanostructures

An overview of earlier work on light emission from SiGe nanostructures in general, including quantum wells, wires, and dots, has been given elsewhere (96). Here, we provide an overview of recent continuous-wave (CW) and time-resolved PL investigations of the recombination dynamics of three-dimensional (3D) $Si/Si_{1-x}Ge_x$ multilayer nanostructures grown by MBE (9,10). The measurements were performed on $Si/Si_{1-x}Ge_x$ nanostructures where a single $Si_{1-x}Ge_x$ nanometer-thick layer (NL) is incorporated into $Si/Si_{0.6}Ge_{0.4}$ 3D-cluster multilayer (CM) structures.

<u>Wavy Nanolayer</u>

Figure 10(a) shows a dark field transmission electron microscope (TEM) image and spatial position of an energy-dispersive X-ray spectroscopy (EDX) scan for the two Si/SiGe cluster layers closest to the Si-substrate of a multilayer structure [see Ref. (9) and references therein]. The EDX data show that the Ge atomic concentration at a Si/SiGe hetero-interface increases from 0 to ~35 at. % within a distance $d \approx 5$ nm [Fig. 10(b)]. In contrast, Fig. 11 shows corresponding data for the two top Si/SiGe clusters layers, where the Ge atomic concentration reaches ~35 at. % within a much shorter distance $d \approx 3$ nm. It is very clear that the topmost Si/SiGe cluster layers in this specially engineered sample (note the increased thickness of the second-to-top layer) have more abrupt Si/SiGe hetero-interfaces compared to those in bottom layers, closer to the Si substrate.

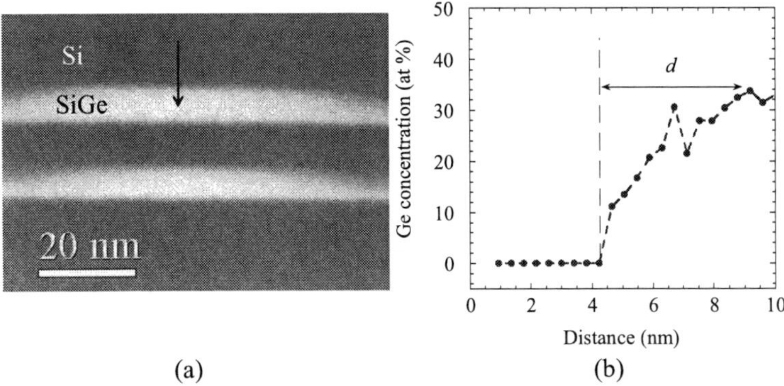

(a)   (b)

Figure 10. (a) A high angle annular dark field scanning transmission electron microscope (HAADF-STEM) micrograph showing the 1st and 2nd bottom SiGe cluster layers (counting from the Si substrate) and EDX scan position (arrow). (b) EDX measured Ge atomic concentration at the 2nd bottom Si/SiGe cluster layer hetero-interface. The interface abruptness d (which is the distance between pure Si and SiGe with the nominal Ge-concentration) is indicated.

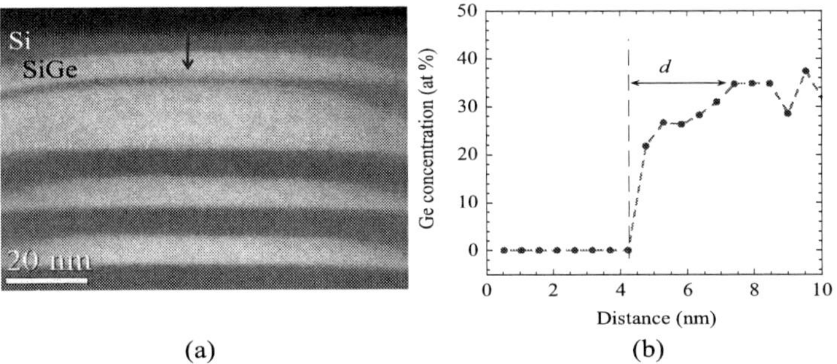

(a)   (b)

Figure 11. (a) A HAADF-STEM micrograph showing top SiGe cluster layers and EDX scan position (arrow). (b) EDX measured Ge atomic concentration at the topmost Si/SiGe cluster layer hetero-interface. The interface abruptness d is indicated.

The PL spectroscopy measurements [see Fig. 12(a)] are consistent with the TEM and EDX data. Using photoexcitation of 405 nm wavelength (which has a penetration depth of ~100 nm), we observe a PL spectrum with a full width at half maximum (FWHM) of ~130 meV. At the same time, using photoexcitation with a wavelength of 325 nm (with an estimated penetration depth of ~10 nm), we find the PL spectrum slightly shifted to lower photon energy and with a significantly reduced FWHM (~100 meV). This result clearly indicates that SiGe cluster composition and Si/SiGe interface abruptness (Figs. 10 and 11) are also reflected in the PL spectra: in top-most SiGe layers, a higher Ge-composition results in the PL peak shifted toward lower photon energy, and the more abrupt Si/SiGe interface is responsible for the reduced FWHM of the PL peak (97).

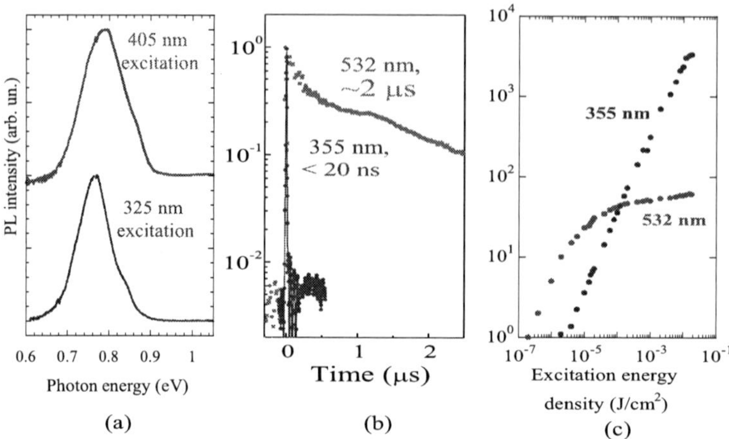

Figure 12. (a) Normalized PL spectra (shifted for clarity) collected at different (indicated) excitation wavelengths. (b) PL decays measured at 0.78 eV under ~6 ns duration pulsed excitation at the wavelengths indicated. (c) PL intensity as a function of excitation energy density at different excitation wavelengths. The sample temperature was 15 K.

Figures 12(b) and (c) compare PL lifetimes and PL intensity as a function of excitation energy density measured under pulsed laser excitation with a pulse duration of ~6 ns. We find that PL excited at 532 nm and recorded at 0.78 eV (which originates from the lowermost Si/SiGe cluster layers) has a lifetime approximately 100 times longer compared to the same PL excited using 355 nm (PL from the topmost Si/SiGe cluster layers). The PL lifetime for 3D SiGe nanostructures is less than 20 ns, and it is independent of excitation energy density until a very high level of excitation. With the PL decays produced under 355 nm excitation where the energy density varied from 10^{-5} to 5×10^{-2} J/cm², no changes in the PL lifetime are found until the Auger limit has been reached, with a carrier concentration approaching $\sim 10^{19}$ cm^{-3} [see Fig. 12(c) and Ref. (98)]. On the contrary, the PL decay under 532 nm excitation quickens while the PL intensity saturates as a function of excitation energy density. Also, we find a dramatic difference in the PL intensity as a function of excitation energy density for excitation at different wavelengths: the PL intensity under 532 nm excitation quickly saturates [in agreement with Refs. (98) and (99)], while under 355 nm excitation the PL intensity is linear versus excitation intensity for many orders of magnitude [Fig. 12(c)].

The results presented here clearly demonstrate that in Si/SiGe 3D nanostructures the PL peaked near 0.78 eV strongly depends on the Si/SiGe hetero-interface abruptness. Our measurements show that in MBE-grown low defect density 3D Si/SiGe nanostructures a transition from pure Si to $Si_{1-x}Ge_x$ with $x \approx 0.3$–0.4 may require an interface width of more than 5 nm. These diffused type II Si/SiGe hetero-interfaces are responsible for significant ($d > 5$ nm) electron-hole spatial separation and slow PL, which cannot compete with Auger recombination even at low excitation intensities. However, by engineering more abrupt Si/SiGe hetero-interfaces with $d \approx 3$ nm, we find a PL lifetime of < 20 ns, which is only a little longer than that found in direct band-gap III-V semiconductors. This extremely fast PL has a quite high quantum efficiency, which, in contrast to previously reported results from similar nanostructures, remains constant over many orders of magnitude of excitation intensity.

<u>Planar Nanolayer</u>

Figure 13(a) shows a TEM image of the sample structure, which consists of a Si substrate (not shown in the figure), a $Si_{1-x}Ge_x$ buffer layer with $x \approx 10$ at. % (seen in part at the bottom of the figure), eight repeats of $Si_{1-x}Ge_x$ cluster (up to 10 nm thick) and Si layer pairs, a single 4-5 nm thick $Si_{1-x}Ge_x$ planar NL followed by a thin Si layer, and a final $Si_{1-x}Ge_x$ cluster layer topped with a Si capping layer (9). The size of the SiGe clusters and thickness of the NL obtained from the TEM measurements is confirmed by EDX measurements, as shown in Fig. 13(b). Also, the EDX scan shows that the $Si_{1-x}Ge_x$ NL composition is relatively uniform with $x \approx$ at. 8%, while in the $Si_{1-x}Ge_x$ clusters x increases from ~5 at. % at the SiGe cluster/Si interface to 40 at. % near the cluster center.

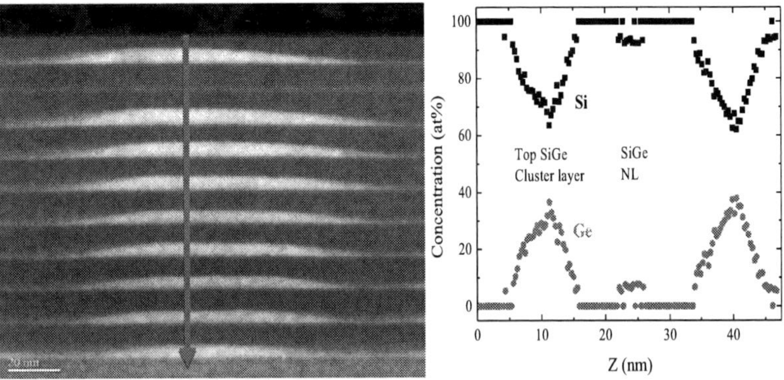

Figure 13. (a) HAADF-STEM image showing the Si/ $Si_{1-x}Ge_x$ NL/CM nanostructure with the very thin $Si_{1-x}Ge_x$ NL fourth from the top (Si) layer, and (b) EDX measured composition of the sample's six topmost layers. The arrow shows the direction of the EDX scan. The scale bar in the TEM image is 20 nm in length.

Figure 14(a) compares the PL dynamics associated with $Si_{1-x}Ge_x$ NL and $Si_{1-x}Ge_x$ clusters using 355-nm wavelength and 6-ns duration pulsed-laser excitation with an energy density of ~50 mJ/cm^2. Nanolayer PL from the sample peaked at 0.92 eV and rises faster than the system time resolution, while CM PL, which peaked at 0.8 eV, has a rise time close to 2–3 μs. Analogous to the wavy layer case, the PL peaked at 0.92 eV is also found to be decaying much faster compared to the PL peaked at 0.8 eV. Non-exponential decays are found for both PL bands of the Si/SiGe CMs. The observed non-exponential PL decays suggest that in both cases the carrier recombination processes are characterized by a time-dependent recombination rate, R_i. Figure 14(b) presents the recombination rate extracted from the PL decay data as a function of time. The recombination rate for the PL band peaked at 0.8 eV is ~10^5–10^4 s^{-1}, and it is in the range of 10^6–10^7 s^{-1} for the 0.92 eV band.

At a higher excitation energy density, the PL peaked at 0.92 eV dominates due to the PL intensity linear dependence versus excitation energy density compared to the sub-linear dependence of the 0.8 eV peaked PL, just like in the wavy layer case. The linear dependence of the 0.92 eV peaked PL intensity versus excitation energy density indicates that the measured recombination rate of 10^6–10^7 s^{-1} is mostly due to radiative recombination. Since radiative recombination competes with Auger recombination, the long-lived PL should saturate sooner compared to the short-lived PL. On the other hand, the 0.8 eV PL rise time as a function of excitation energy density shows different behavior at low and high excitation energy densities (10). The observed PL rise time of ~2–3 μs is much longer than the laser pulse (~6 ns). This unusually long PL rise time could be associated with an Auger-assisted carrier spatial redistribution in Si/SiGe nanostructures known as the Auger fountain (100). The temperature dependence of the PL rise time at high excitation density (~50 mJ/cm^2) also confirms that the Auger fountain could be responsible for the unusual PL dynamics (101).

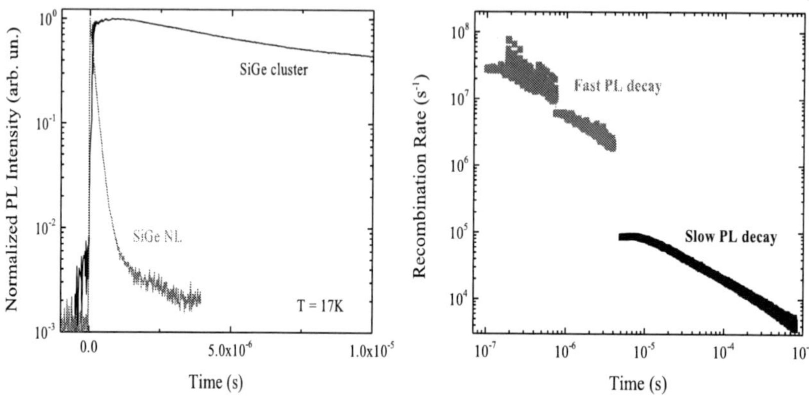

Figure 14. (a) Low temperature (17 K) time-resolved PL decays under pulsed excitation energy density of 50 mJ/cm^2 recorded at photon energies associated with Si$_{1-x}$Ge$_x$ NL PL (~0.92 eV) and Si$_{1-x}$Ge$_x$ cluster PL (~0.8 eV). (b) Carrier recombination rate as a function of time calculated using the PL decay data. The fast and slow PL decays are from the NL and clusters, respectively.

Non-exponential PL decays have been reported previously in Si/SiGe NSs, and they were fitted variously by a stretched exponential function $\exp[(-t/\tau)^{\beta}]$, a power function $1/(1 + at)^m$, or multiple exponential decays (98,102,103); however, the underlying physical mechanism involved has not been identified. It has been pointed that the stretched exponential PL decay is observed in a wide variety of systems and although it provides a good empirical fit, it most likely has no fundamental significance (104). Instead, we extract the recombination rate directly from the PL decays without using any assumption or a specific model. Figure 14(b) shows that initially both PL bands have almost time-independent recombination rates with corresponding single-exponential decays of ~3×10^7 s^{-1} for the PL band peaked at 0.92 eV and ~9×10^4 s^{-1} for the 0.8 eV PL band (most likely due to the limited system time resolution). As time increases, the recombination rate decreases, and $R_i(t) \sim t^{\alpha}$ with $\alpha \approx 0.82$ for the PL band peaked at 0.92 eV and $\alpha \approx 0.67$ for the PL band peaked at 0.8 eV.

We assume that the holes are localized within SiGe and the electrons are located in Si, which is due to the type II energy band alignment at the Si/ $Si_{1-x}Ge_x$ hetero-interface. Just as in the donor-acceptor pair (DAP) recombination model (105), we explain the electron-hole time-dependent recombination rate by assuming that it depends on the average distance separating electrons and holes, a_{e-h}. The recombination-rate distance dependence is expressed as

$$R(a) = R_0 \exp(-\frac{a_{e-h}}{a_0})$$ [2]

where R_0 and a_{e-h} are the maximum recombination rate [~ 10^8 s^{-1}, see Fig. 14(b)] and a minimal radius of the localized exciton at the Si/ $Si_{1-x}Ge_x$ hetero-interface (~1.5 nm), respectively. In the $Si_{1-x}Ge_x$ NL with $x \sim 8$ at. %, we find $a_{e-h} \leq 5$ nm (which is comparable to the thickness of the $Si_{1-x}Ge_x$ NL) while in $Si_{1-x}Ge_x$ CMs with $0 \leq x \leq 40$ at. %, we find 9 nm $< a_{e-h} <$ 14 nm (Fig. 15). These results are in good agreement with the TEM and EDX data (see Fig. 13).

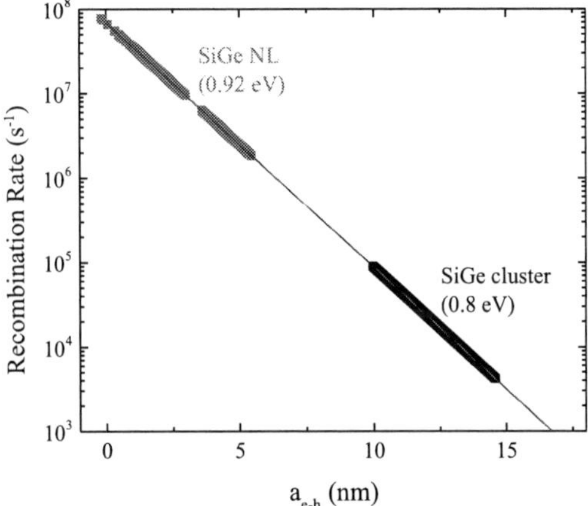

Figure 15. Carrier recombination rates (dots) extracted from the experimental data as a function of the distance between electrons and holes for photon detection energies associated with $Si_{1-x}Ge_x$ NL PL (~0.92 eV) and $Si_{1-x}Ge_x$ cluster PL (~0.8 eV). The solid line is the theoretically calculated electron-hole recombination rate.

In summary, these experiments have shown that two quite different $Si_{1-x}Ge_x$ NLs incorporated into a $Si_{0.6}Ge_{0.4}$ CM structure exhibit an intense PL signal with a characteristic decay time as much as 1000 times shorter than that observed in Si/SiGe CMs. The experimentally observed non-exponential PL decays in Si/SiGe nanostructures can be explained as being due to variations of the distances separating electrons and holes at the Si/SiGe hetero-interface. It is seen that an abrupt Si/$Si_{1-x}Ge_x$ hetero-interface reduces the carrier radiative recombination lifetime and increases the PL quantum efficiency, making these $Si_{1-x}Ge_x$ NL/CM nanostructures promising candidates for applications in CMOS compatible light-emitting devices.

SiGe Nanowire Arrays

Semiconductor nanowires (NWs) are thought of as promising building blocks for opto-electronic devices that exploit their novel electronic band structures generated by two-dimensional (2D) quantum confinement in conjunction with their associated optical properties (106–108). However, in order to fully implement these new properties, strict control is needed over the NW location, uniformity, composition, and size. Many of the existing NW growth methods have led to NWs possessing non-uniform diameters and lengths and that are haphazardly oriented and randomly positioned (109). An efficient and simple electrochemical process has been developed that combines focused-ion-beam (FIB) lithography and galvanic reaction to selectively synthesize gold nanoparticles in well-defined locations that are subsequently used for the MBE growth of ordered $Si_{1-x}Ge_x$ NW arrays with predefined NW positions and diameters (11), as shown for example in Fig. 16. Here we summarize the optical properties of such MBE-grown well-ordered NWs.

Three NW samples were prepared for this study (11): Sample (A), where the NWs are grown randomly across the Si substrate; sample (B), where the nanowires decorate the edges of 400x400 μm^2 boxes; and sample (C), where the NWs fill 400x400 μm^2 boxes in ordered arrays, as described above. These samples have NWs that have a nominal Ge concentration of x = 0.15 and are 200 nm in diameter and 200 nm long, with a morphology similar to the Si NWs shown in Fig. 16.

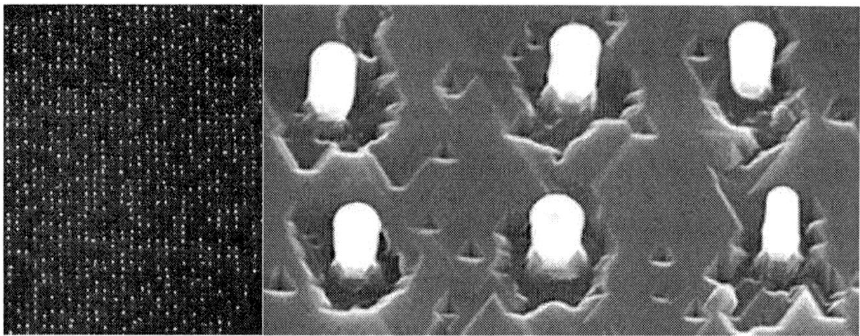

Figure 16. Scanning electron microscope (SEM) images of the ordered arrays of Si NWs showing (left) a NW array and (right) individual 200 nm long NWs.

The temperature dependence of the PL spectrum obtained from sample (C) is shown in Fig. 17. PL spectra with similar temperature dependences were obtained from the other samples. Figure 17 shows that the NW spectral region of interest (from approximately 950 to 1050 meV) is dominated by the boron ($\sim 10^{17}$ cm^{-3}) doped Si-substrate phonon-replica spectrum at the lowest temperatures (6 and 10 K). Upon increasing the sample temperature up to 20 K, the Si-substrate PL becomes sufficiently quenched from the increasing dissociation of multiple-donor bound excitons within the substrate (110) such that the underlying NW PL is more readily seen. By 25 K, only one sharp line at 1092.5 meV due to the Si substrate remains. Although their spectra were quite similar, the overall intensity of the NW PL varied from sample to sample; sample (C), with a higher density of NWs distributed within the array, was the strongest, while sample (A) with a random distribution of NWs was the weakest.

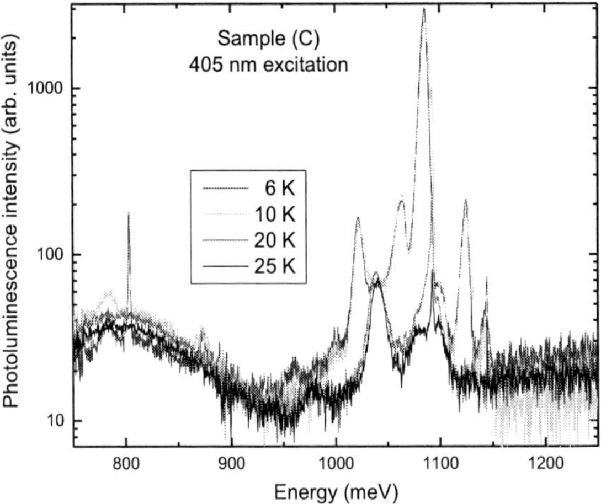

Figure 17. Temperature dependence of the instrument-response-corrected PL spectrum obtained from sample (C) with excitation at 405 nm.

The NW PL is strong considering that the volume of NW material is so small (each wire has a volume of 0.006 μm^3) and thus the carriers have to be recombining efficiently within the wires to produce this level of light emission at the SiGe-alloy band gap energy. We know from earlier studies of SiGe etched wires (111) and dots (96) that just the spatial confinement of carriers is sufficient to produce the readily-observed PL found here. This implies that the carriers in these samples are not being lost in large numbers to the substrate or recombining in large quantities at defects inside or on the surface of the NWs. The wires are too large in diameter for a quantum confinement induced energy shift in the band gap, but the phonon energies could be affected slightly by confinement and surface effects (112). The wires are grown free-standing and thus there should be no internal strain (i.e., bulk-like energy values should be observed).

Further details about the PL spectra can be obtained from spectral curve resolving, Such an analysis using a Gaussian line shape revealed that there are four major features in the energy range of interest (see Fig. 18 for typical results obtained at 25 K). In order of increasing energy, they are readily assigned to the NW free-exciton transverse-optic (TO) Si-Si vibrational mode, NW free-exciton transverse-acoustic (TA) phonon, Si-substrate-bound-exciton TO phonon and NW free-exciton no-phonon (NP) lines, respectively (90,110). The amplitude ratio for the TO/TA peaks is much the same in the three samples, as would be expected if both lines arose just from the NWs and that the NWs were of similar composition in all samples. The fitted frequencies and line widths for the respective NW and Si TO mode lines are the same within error. Interestingly, the free-exciton NP line intensity relative to its phonon replicas (TA and TO lines) in these NWs is much more intense than that found in bulk Si, but is somewhat lower compared to what is observed in bulk alloy material of a similar composition (90).

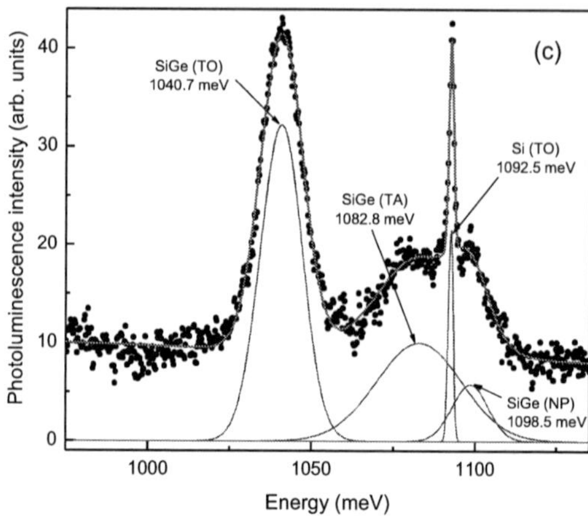

Figure 18. Curve-resolved PL spectrum of sample (C) at 25 K obtained with 405 nm excitation. The solid line shows the overall fit to the PL data, while the three SiGe NW component lines are shown beneath the fitted spectrum. The very sharp line at 1092.5 meV arises from the Si substrate.

From the fits to the sample with the strongest PL [sample (C)], the NW TA and TO phonon energies are found to be 15.7 ± 1.8 and 57.8 ± 0.6 meV, respectively, which agree very well with the values expected for bulk $Si_{1-x}Ge_x$ with $x = 0.15$ of 18 ± 1 and 58 meV, respectively (90). The measured NW free-exciton NP energy of 1099 meV would indicate a Ge concentration of $x = 0.14$ (giving an X-point energy gap of 1099 meV in bulk SiGe) (90). Thus the positions in energy of the NP peak and those of the accompanying phonon replicas independently confirm the alloy concentration as being $x = 0.15 \pm 0.01$.

In summary, the readily-observed PL arising from the SiGe NWs indicates they are clean (i.e., contain few growth defects and impurities) and are electrically isolated from the substrate. They are not strained to any significant extent and x for these samples is confirmed from the PL to be 0.15. These NWs with their well-controlled position, composition, and size and their efficient luminescence exhibit relevant features that are a significant improvement in quality over those produced by other vapor-solid-solid growth methods and that could be useful for applications in optoelectronic nanodevices. However, their mass production in current CMOS production lines would be problematic.

Optical Emission from Germanium Nanocrystals

Previously, a very intense, low temperature PL with a long lifetime has been observed in numerous samples of MBE-grown $Si_{1-x}Ge_x$ (SiGe) epitaxial layers with x ranging from 0.05 to 0.53 (113). Efficiencies in the 5% range – unheard of for group IV materials – were observed. This PL was neither defect nor dislocation related, and was suspected to be due to carrier localization effects. We have shown recently (114) that the PL is characteristic of self-assembled Ge nanocrystals (NCs) contained within the alloy epilayers through an investigation of its concentration dependence, both experimentally and theoretically. The occurrence of these strained Ge NCs is confirmed through Raman spectroscopy measurements.

As can be seen in Fig. 19, the PL that we attribute to Ge NCs consists of a broad peak with an asymmetry to low photon energies. This peak displays little variation in shape with Ge-fraction and tracks the band gap (BG) variation, but is ~100 meV below the indirect SiGe BG. If the material were Si or SiGe, the width of the peak at ~50 meV would be too small for it to be due to a no-phonon (NP) line with its transverse-optic (TO) phonon replica, as the NP-TO spacing is about 58 meV for those materials. Figure 19 shows that for higher Ge-fractions the PL is emitted at energies significantly below those for bulk Ge, with its indirect BG of 744 meV at low temperatures. The origin of this broad, intense PL peak has remained unexplained, until now. In Fig. 19 we display low temperature PL from three SiGe samples and bulk Ge. For the spectrum in trace (a), the result is also illustrated with the results of curve fitting using two Gaussian peaks for the broad PL peak.

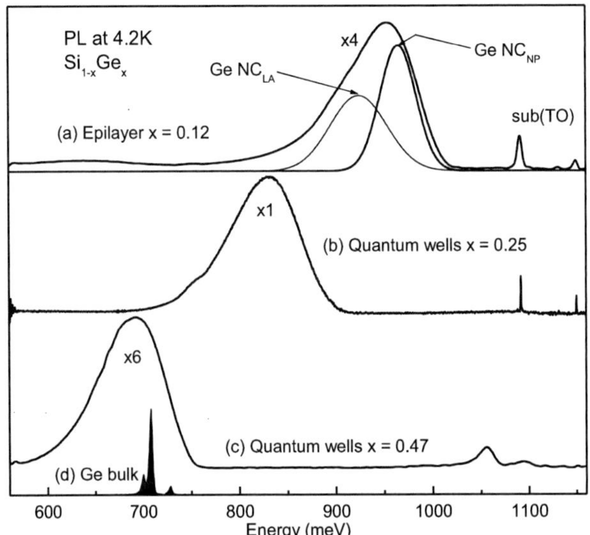

Figure 19. Low temperature PL from (a)-(c) three $Si_{1-x}Ge_x$ samples and (d) bulk Ge. For the spectrum in trace (a), the result is also illustrated with curve fitting using two Gaussian peaks for the broad PL peak.

As is shown in Fig. 19(a), the broad peak's width and asymmetric shape can be curve-resolved into two symmetric peaks, separated by ~35 meV, i.e., very near the momentum conserving TO phonon energy for Ge. The NP peak is wide (25-45 meV) due to NC confinement variations arising from size variability and to alloy disorder broadening in the SiGe. In Ge-dots, we expect two main PL peaks; a NP line and a longitudinal-acoustic (LA) phonon replica line, separated by about 28 meV. So the NP-phonon replica energies provided by curve resolving are relatively close to that (28 meV) for the most intense (LA) phonon replica for Ge, but differ very significantly from the corresponding energy of the strongest (TO) phonon replica for Si and SiGe, which, in both cases, is 58 meV. We also notice inhomogeneous broadening due to size effects and that the NP peak is relatively large, as would occur for a high degree of carrier localization in Ge NCs.

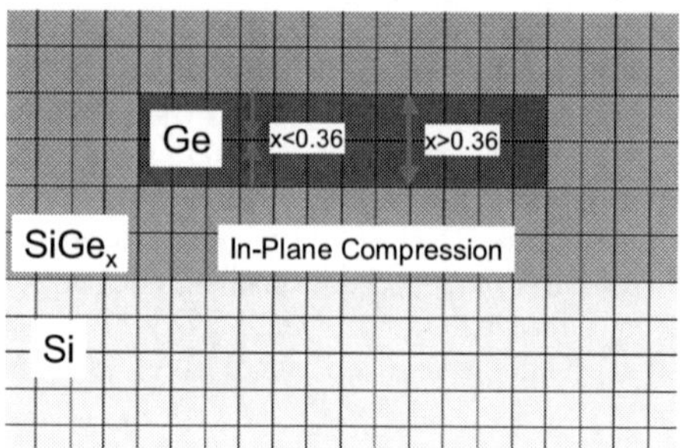

Figure 20. Cross-sectional schematic for the x-z plane of a lattice matched Ge NC within a $Si_{1-x}Ge_x$ epilayer on Si(001).

An examination of the available experimental evidence indicates that a possible cause of the broad peak is imbedded Ge NCs within the SiGe layers (see Fig. 20). The formation of such NCs was shown in TEM studies (113) and by Raman spectroscopy (114). Referring to Fig. 20, we see that ideal epitaxial growth means that the SiGe epilayers and the Ge NCs be lattice matched to Si (001) in the x-y plane, so that both the SiGe and NCs are under compression in that plane. However, the growth of SiGe is unconstrained in the vertical (z) direction. As is well known, the SiGe epilayer is under tensile strain in this direction, leading to a vertical lattice constant that increases with Ge fraction and is larger than that for unstrained SiGe. Here the volume of the unit cell for the epitaxial SiGe layer is assumed to be the same as that for unstrained cubic (bulk) SiGe. An important assumption is that the lattice constant of the Ge NCs matches that of the SiGe in all three directions, i.e., including vertically. For relatively dilute SiGe, the Ge NCs are under compression vertically, but for increasing Ge-content in the SiGe epilayer the vertical lattice constant of the strained SiGe eventually exceeds that of bulk Ge. At the point where the vertical strain in the Ge NC first becomes tensile, the Ge-fraction in the SiGe is 0.36. With these constraints we can write down an expression for the strain in the z direction within the Ge NCs as:

$$\varepsilon_z = (c^{Ge} - c^{SiGez})/\, c^{Ge} \tag{3}$$

where c^{Ge} is the bulk Ge lattice constant (5.658 Å) and c^{SiGez} is the strained lattice constant in the vertical direction for tetragonally distorted SiGe material, which is lattice matched to Si in the horizontal plane. Note that the strain value is negative for compression and positive for tension. With increasing x, this lattice constant varies from 5.431 to 6.141 Å as given by the following expression:

$$c^{SiGez} = (c^{SiGeu})^3/(c^{Si})^2 \tag{4}$$

where c^{Si} is the lattice constant for Si (5.431 Å). Here c^{SiGeu} is the lattice constant for cubic (unstrained) SiGe given by:

$$c^{SiGeu} = c^{Si} + c_1 x + c_2 x^2 \qquad [5]$$

where x is the Ge fraction. In this quadratic equation, the symbols c_1 and c_2 represent constants equal to 0.2 and 0.027, respectively (115). Strain versus Ge-fraction is thus easily derived for the vertical (z) direction as shown by the horizontal scale at the bottom of the graph in Fig. 21.

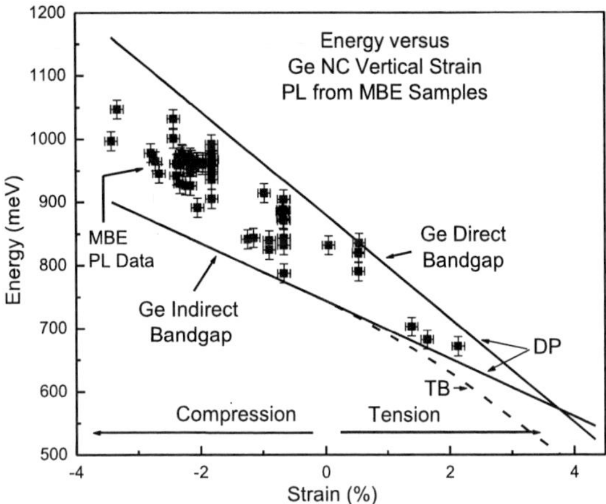

Figure 21. No-phonon line PL peak energy and bandgap (direct and indirect) energies of Ge versus Ge NC vertical strain. For the indirect gap under tensile strain, the energy has been calculated using the deformation potential (DP) and tight bonding (TB) theories.

Strain has a relatively large effect on the material bandgap energy, with compression generally increasing this energy while tension reduces it. A relatively simple method for calculating this effect is to use deformation potential (DP) theory, which employs linear relationships between strain and the direct and indirect bandgaps, as per:

$$E_{BG} = A + B\varepsilon_z \qquad [6]$$

Here, for strain expressed in per cent, the intercept and slope in the linear equation are 880 and -79.3 meV for the direct bandgap and 740 and -45.5 meV for the indirect one at low temperatures (116). For example, for a tensile strain of 2 %, the direct BG energy is 721 meV and the indirect BG is 649 meV.

From DP theory, as the strain becomes more strongly tensile both the direct and indirect gap energies decline, with the direct energy decreasing more rapidly than the indirect one. As shown in Fig. 21, the direct gap energy crosses over the indirect gap energy at a tensile uniaxial strain of 4 %, resulting in Ge becoming a direct gap semiconductor, a highly desirable outcome. Our results with PL point the way to this transition point, but the maximum vertical tensile strain for the present Ge NCs is not much greater than 2 %. Nonetheless, the fact that we see PL below the indirect BG of bulk Ge is explained by these particular Ge NCs being under tensile strain vertically, which reduces their Ge BG. There are other nonlinear models for the bandgap energy such as the tight binding (TB) model (116), for which Fig. 21 contains the indirect bandgap calculation under tensile strain.

For the bandgap curves calculated versus strain in Fig. 21 the effects of quantum confinement have not been included. The data points were obtained by curve resolving the PL of the MBE-grown SiGe samples. In Fig. 21 the points are Ge NC NP PL peak energies and the solid lines are the direct and indirect NC BGs (Equation 6). The NP experimental peak energies have been obtained from the broad PL distributions by fitting Gaussian functions to the observed PL spectra. In Fig. 21 we note that the data falls generally above the indirect energy BG for uniaxially strained Ge, a differential which is most likely due to quantum confinement, as will be discussed next. This is a difference that appears to decrease as compression decreases, although there are only a few points for the larger tensile strains. The energy uncertainties in the data points are those obtained from curve resolving the PL peaks. The strain uncertainties for these points are derived from the X-ray diffraction measurements of the composition and thickness of the SiGe layers.

One point for discussion is whether or not quantum confinement for a reasonable range of NC sizes can account for the PL being at the energies above the strained bulk Ge indirect BG, as we see in Fig. 21. This difference is in the 50 to 150 meV range with an average of 103.4 meV. So we are asking whether such values are reasonable for quantum confinement effects in the Ge NCs. With the simple model described above we can provide an answer by calculating the size of the confinement effect versus NC size, quantum well thickness, and composition. In general it was found that the variation of NC confinement energy with thickness of the host SiGe was not that large (< 5 meV) and the variation with Ge-fraction although larger was not a major contributor to the difference. For example, the NC confinement shift for a 1.5 nm NC in a 5 nm thick well was 110 meV for a Ge-fraction of 0.15 and 130 meV for a fraction of 0.50, everything else being equal. However, NC sizes in the range from 1 to 3 nm – corresponding to confinement energies from 120 to 50 meV – can indeed account for all the blue shifts from the predicted Ge NC BG that we see in the PL energies. This means that confinement shift provides a likely explanation for the difference seen between the measured PL energy and the NC bandgap and we estimate that the NC size varies from about 1 nm for a compressive strain of 4 % to 3 nm for tensile strain of 2 %.

The Raman spectroscopy results obtained from a representative set of samples were curve-fitted to obtain the various mode frequencies (114). The concentration dependence of the Ge-Ge (arising from the $Si_{1-x}Ge_x$ alloy layer) and Ge (arising from the Ge NCs) mode frequencies are shown in Fig. 22. These data are consistent with the results of an earlier study (117). The existence of a sharp, although weak, Ge line in all spectra confirms the presence of Ge NCs in all of the samples studied by Raman spectroscopy. The approximately linear shift in frequency with x is a result of the strain in the alloy layer in the sample growth (z) direction, which is the primary direction of phonon propagation sampled in the Raman back-scattering experiment.

From Chen et al. (118), the Ge-Ge phonon mode frequency $\omega_{Ge\text{-}Ge}$ in $Si_{1-x}Ge_x$, as seen by Raman scattering, depends on the germanium fraction x and strain ε as follows:

$$\omega_{Ge\text{-}Ge} = 282.5 + 16x - 385\varepsilon. \qquad [7]$$

For the Ge NCs, x is equal to unity. In the vertical direction the relevant strain, ε_z, for the NCs is given by Equation 3, leading to the red line in Fig. 22, in which the phonon frequency declines from 312 to 290 cm^{-1} as x increases from 0.1 to 0.55. In the x-y plane the NC strain is compressive and constant at -4.18 %, because the NCs are lattice matched to Si in this plane. Hence in the x-y plane, the NC Raman frequency is not expected to vary with Ge fraction in the surrounding SiGe, as shown by the green horizontal line at a Raman frequency of 314.6 cm^{-1} (calculated as per Equation 7) in Fig. 22.

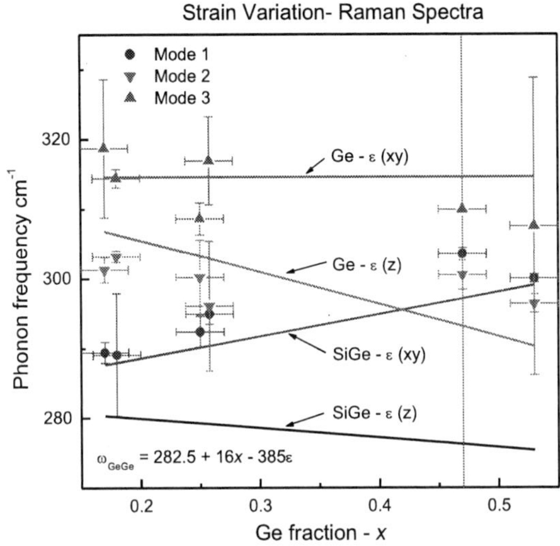

Figure 22. Concentration dependence of the zone-center Ge-Ge and Ge phonon mode frequencies, obtained from Raman scattering. For example, the blue points ($\bullet$) correspond to a Ge-Ge mode in the Si$_{1-x}$Ge$_x$.

In the x-y plane for the SiGe layers, the increasingly compressive strain causes the Ge-Ge Raman mode frequency to rise with Ge-fraction, as is apparent for the blue line in Fig. 22 for which the frequency goes up from 285 to 300 cm^{-1} when the Ge-fraction changes from 0.1 to 0.55. In the vertical (z) direction the strain is tensile in the SiGe layers and increases with Ge-fraction, which progressively suppresses the Ge-Ge phonon mode frequency, as seen in the black line below 280 cm^{-1} in Fig. 22. The theoretical predictions are in general agreement with experiment, as can be seen in Fig. 22, which confirms our assignments and the presence of Ge NCs.

In conclusion, our modelling of the experimental results for the concentration dependence of the PL from strained Si$_{1-x}$Ge$_x$ epilayers for $0.05 < x < 0.55$ has shown that the intense PL at lower energies arises from imbedded Ge NCs. Such structures show promise for use as light emitters in device applications if appropriate growths at high x values can be engineered to introduce the predicted transition to a direct band gap within strained Ge NCs for x-values greater than 0.65.

Other Nanostructures

A general overview of experimental and theoretical research into the optical properties of both crystalline and amorphous nanostructures formed from both Si and Ge has been given recently by Barbagiovanni *et al.* (112). Several of the most-recently-published papers providing new information in this field of research are profiled next.

<u>Germanium Light Sources</u>

Germanium nanostructures have shown considerable promise for light emitting applications, as the indirect band gap of Ge can be converted to a direct one through a combination of applying tensile strain and appropriate doping at a high level [see, for example, the review of this field in Ref. (6)]. Although laser devices based on Ge have been demonstrated some time ago, at present there is no ability to produce such devices using complementary-metal-oxide-semiconductor (CMOS)-compatible processes (6). New information on the optical properties of strained Ge nanostructures has been obtained recently from fresh investigations of the intense broad PL produced by self-assembly during molecular beam epitaxy growth of strained $Si_{1-x}Ge_x$ alloy layers on Si (113,114). A theory based on modelling the Ge nanostructures as quantum wires reproduces the observed energy shift of the PL and confirms that embedded Ge nanocrystals give rise to this light emission via carrier localization in three dimensions. By increasing the Ge-concentration of the alloy epilayer(s), the strain felt by the crystalline Ge nanostructures can be varied from compressive to tensile with the crossover occurring at $x = 0.36$. Such encapsulated Ge nanostructures are calculated to yield a direct band gap for $x \geq 0.7$ and thus are good candidates for the production of a laser at near infrared wavelengths.

A different approach to obtaining bright light from Ge has been adopted by Moritz Brehm and co-workers (4,119–121). They have explored the properties of disordered-Ge quantum dots and found bright light emission at room temperature and even stimulated emission, much as was previously observed from disordered and crystalline Si quantum wells (51,86,122), but with the added advantage that the emission was at a desirable near infrared wavelength in the range 1.3–1.55 μm rather than in the red for Si quantum wells. They have also demonstrated a light emitting diode based on their defect enhanced Ge quantum dots (121), which augers well for the integration of these light emitting nanostructures into conventional CMOS production lines.

Another novel approach towards improving the intensity of light emission from Ge dots is reported by Rutckaia *et al.* (123). Their method takes advantage of Mie resonances in specially constructed Si-based nanostructures, which involves coupling between the self-assembled Ge quantum dots and Si nanodisks. The near-infrared PL from an embedded dot is considerably enhanced due to a good spatial overlap between its physical position and the electric field of a Mie mode. Construction of structures containing nanodisk trimers resulted in a further 10-fold enhancement of the light emission. Such resonant dielectric nanostructures provide a new CMOS-compatible platform for the manipulation of light at the nanoscale.

In another ground-breaking piece of research, Amollo *et al.* (124) have explored the structural, optical, electronic and magnetic properties of reduced graphene oxide–Ge quantum dot nanocomposites. Graphene has been of great interest recently for its applications in nanoelectronics and optoelectronics and this paper advances its use in developing nanocomposites for such applications. The functionalization of graphene was achieved via a single-step microwave-assisted solvothermal synthesis to produce the require nanocomposite structure containing quantum confined Ge ad-atoms on reduced graphene oxide sheets. The highly crystalline dots were spherical in shape and had diameters ranging from 1.6–9 nm. Optical spectroscopy revealed a clear dot size dependence of the light absorption in the ultraviolet region, strong and sharp PL at 380 nm in the ultraviolet, and a moderate intensity and broad emission at 450 nm with a tail extending across the visible wavelength region. These results are preliminary, but point to interesting possibilities in optoelectronics and the need for further research. From a device point of view, these composite nanostructures are readily dispersible in a polar solvent, which makes them appropriate for cost effective processing in solution.

The complexities of charge carrier behavior (creation, exchange, and recombination) in Ge quantum dots (dimensions: 14 nm across the base and 2.7 nm high) immersed in Si have been revealed in detail from a mid-infrared optical absorption study carried out by Vorobjev *et al.* (125). By measuring the variation in the optical transmission spectrum of the nanostructures under interband illumination, the authors have found that there is an increase of absorption at the long-wavelength edge of the spectrum, which ranges from 2–5 μm, and a decrease of absorption at the short-wavelength edge. The increase of absorption is observed for light polarized along the growth direction and is related to the contribution of optical transitions involving non-equilibrium holes arising from the quantum dot ground state. The decrease of absorption at shorter wavelengths is observed for the same light polarization. This effect is proposed to be caused by the suppression of interband-like optical transitions between bound states in the valence and conduction bands and, interestingly, is essentially a *dynamic* analogue of the Burstein–Moss effect in bulk semiconductors. From an application point of view, it is useful to know that the optical transmittance variance related to non-equilibrium charge carriers becomes insignificant for temperatures greater than 250 K.

<u>Silicon Light Sources</u>

Silicon nanostructures continue to be subjects of research with regards to improving their light emitting properties. Recent work by Vieira *et al.* (126) has extended earlier work on Si/SiO_2 multilayers (107) to investigate alloys of Ge with Si in the quantum wells, and, after high temperature annealing, thereby produce SiGe nanocrystals. By varying the alloy layer thickness from 2–3.5 nm, they were able to change the disposition of the alloy layer from isolated SiGe nanocrystals of 3–8 nm in size to a continuous layer of crystalline SiGe. However, only weak PL in the range 0.7–0.9 eV was observed from the nanocrystals at low temperature. The emission exhibited some structure that was associated with the presence of interface states between the crystalline nanostructures and the amorphous matrix surrounding them.

Investigations of the optical absorption of Si nanocrystals in single and multilayer structures where the nanocrystals were confined between variable thickness SiO_2 barrier layers have been carried out by Greben *et al.* (127). The absorption cross-section was determined as a function of barrier thickness by using a PL modulation method. A large variation of the absorption strength with barrier layer thickness was observed and explained as a consequence of the efficiency of energy transfer between the layers of Si nanocrystals and/or defect population states. Such information is important when designing structures containing Si nanocrystals for use in next-generation Si-photonics and Si voltaic devices.

Recent theoretical (128) and experimental (128,129) work on the optical properties of Si nanocrystals has shed more light on the emission mechanisms involved in earlier observations reported in 2010 of a red spectral shift and enhanced quantum efficiency in phonon-free PL from silicon nanocrystals (130). In the News and Views Section of the same issue of *Nature Nanotechnology* as Ref. (130), Kovalev wrote "Can silicon ever be a true direct-bandgap semiconductor? The first observation of a new, short-lived PL band from silicon nanocrystals offers fresh hope." (131), which raised expectations for new applications of Si nanocrystals within Si optoelectronics. The new theoretical and experimental work of Ref. (128), however, has cast doubt on the interpretation of the original experimental work (130). Luo *et al.* (128) suggest that state filling assisted by strong pumping with subsequent fast recombination of multi-excitons could be responsible for the red-shifted emission, as higher states are emptied faster for smaller nanocrystals. Consequently, the very fast recombination observed in Ref. (130) could be attributed to multi-exciton or Auger recombination, and thus is not the highly-desired direct-band transition. In reply, de Boer *et al.* (129), after reviewing a number of different aspects of the work of Luo *et al.* (128) and other related works, assert that Luo *et al.* have not untangled the mystery of source of the red-shift in band energy. They think that the results of Luo *et al.*'s modelling actually support their previous assignment of the fast PL band as arising from the Γ-valley character admixture within Si-nanocrystals into states with energies below the bulk-Si direct band gap. Clearly, more experimental and theoretical work on this topic is necessary before proceeding with any technological applications requiring use of a direct band gap in Si nanocrystals.

Conclusion

The development of a light emitter compatible with Si based CMOS circuit technology and fast optical interconnects is very important for the upcoming generations of microprocessors and computers. Self-assembled Si, Ge, and SiGe nanostructures with light emission in the all-important optical communication wavelength range of 1.3–1.55 μm are compatible with conventional CMOS processes and, based on present results obtained with specially engineered structures, are also able to produce the requisite fast emission times. The wide variety of new bright sources of PL considered here are promising, but not yet ready for applications, and require further research and development work. At present, it appears that strained Ge nanostructures are the most promising ones for the near-infrared lasers needed urgently for their compatibility with fiber optics.

Acknowledgments

This paper has been based on material covered in my Gordon E. Moore Medal Address at the ECS Spring 2019 Meeting held in Dallas, Texas. I am very grateful to the ECS for recognizing our work with this award, to the National Research Council Canada for their foresight in allowing this research to proceed for more than three decades, and to my many colleagues there for their support. Some of the results summarized here have been published elsewhere in greater detail in Refs. (8–12,89,114). This particular work has been carried out in extensive collaborations here at the National Research Council, Ottawa, Canada with primarily Drs. J.-M. Baribeau, J. P. McCaffrey, N. L. Rowell, and X. Wu; and with the research groups of Professor L. Tsybeskov, New Jersey Institute of Technology, Newark, New Jersey, USA and Professor I. Berbezier, Institut Matériaux Microélectronique Nanosciences de Provence, Marseille, France. I am grateful for the many contributions of these and my many other colleagues to this joint work.

References

1. D. J. Lockwood, Ed., *Light Emission in Silicon*, Academic Press, New York (1998).
2. J.-M. Baribeau, X. Wu, N.L. Rowell, and D.J. Lockwood, *J. Phys.: Condens. Matter* **18**, R139 (2006).
3. J.-N. Aqua, I. Berbezier, L. Favre, T. Frisch, and A. Ronda, *Phys. Reports* **522**(2), 59 (2013).
4. M. Brehm and M. Grydlik, *Nanotechnology* **28**, 392001 (2017).
5. L. Pavesi and D. J. Lockwood, Eds., *Silicon Photonics*, Springer, Berlin (2004).
6. S. Saito, A. Z. Al-Attili, K. Oda, and Y. Ishikawa, *Semicond. Sci. Technol.* **31**, 043002 (2016).
7. D. J. Lockwood and L. Tsybeskov, *IEEE J. Sel. Top. Quantum Electronics* **20**(4), 8200807 (2014).
8. D. J. Lockwood, N. L. Rowell, A. Gouyé, L. Favre, A. Ronda, and I. Berbezier, *ECS Trans.* **61**(5), 31 (2014).
9. S. A. Mala, L. Tsybeskov D. J. Lockwood, X. Wu, and J.-M. Baribeau, *Appl. Phys. Lett.* **103**, 033\103 (2013).
10. S. A. Mala, L. Tsybeskov, D. J. Lockwood, X. Wu, and J.-M. Baribeau, *Physica B* **453**, 29 (2014).
11. A. Benkouider, A. Ronda, A. Gouye, C. Herrier, L. Favre, D. J. Lockwood, N. L. Rowell, A. Delobbe, P. Sudraud, and I. Berbezier, *Nanotechnology* **25**(33), 335303 (2014).
12. D. J. Lockwood, N. L. Rowell, A. Benkouider, A. Ronda, L. Favre, and I. Berbezier, *Beilstein J. Nanotechnol.* **2014**(5), 2498 (2014).
13. A. Uhlir, *Bell Syst. Tech. J.* **35**, 333 (1956).
14. D. R. Turner, *J. Electrochem. Soc.* **105**, 402 (1958).
15. T. Unagami and M. Seki, *J. Electrochem. Soc.* **125**, 1339 (1978).
16. M. I. J. Beale, J. D. Benjamin, M. J. Uren, N. G. Chew, and A.G. Cullis, *J. Cryst. Growth* **73**, 622 (1985).
17. G. Bomchil, A. Halimaoui, and R. Herino, *Appl. Surf. Sci.* **41-42**, 604 (1990).
18. M. I. J. Beale, N. G. Chew, M. J. Uren, A. G. Cullis, and J. D. Benjamin, *Appl. Phys. Lett.* **46**, 86 (1985).
19. K. Barla, R. Herino, G. Bomchil, J. C. Pfister, and A. Freund, *J. Cryst. Growth* **69**, 726 (1984).
20. Y. Watanabe, Y. Arita, T. Yokoyama, and Y. Igarashi, *J. Electrochem. Soc.* **122**, 1351 (1975).
21. K. Imai, *Sol. State Electron.* **24**, 159 (1981).
22. C. Pickering, M. I. J. Beale, D. J. Robbins, P. J. Pearson, and R. Greef, *J. Phys. C* **17**, 6535 (1984).
23. L.T. Canham, *Phys. World* **5**(3), 41 (1992).

24. L.T. Canham. *Appl. Phys. Lett.* **57**, 1046 (1990).
25. A.G. Cullis and L.T. Canham, *Nature* **353**, 335 (1991).
26. V. Lehmann and U. Gösele, *Appl. Phys. Lett.* **58** 856 (1991).
27. D. J. Lockwood, *Solid State Commun.* **92**, 101 (1994).
28. D. J. Lockwood, G. C. Aers, L. B. Allard, B. Bryskiewicz, S. Charbonneau, D. C. Houghton, J. P. McCaffrey, and A. Wang, *Can. J. Phys.* **70**, 1184 (1992).
29. D. J. Lockwood, A. Wang, and B. Bryskiewicz, *Solid State Commun.* **89**, 587 (1994).
30. D.J. Lockwood, A.G. Wang and B. Bryskiewicz, in *Quantum Confinement: Physics and Applications*, M. Cahay, S. Bandyopadhyay, J. P. Leburton, A. W. Kleinsasser, and M. A. Osman, Eds., p. 222, Electrochem. Soc., Pennington, NJ, (1994).
31. R. Siegele, H. K. Haugen, D. J. Lockwood, B. Bryskiewicz, J. F. Forster, and H. R. Andrews, *Solid State Commun.* **93**, 833 (1995).
32. D. J. Lockwood and A.G. Wang, *Solid State Commun.* **94**, 905 (1995).
33. P. Schmuki, L. E. Erickson, and D. J. Lockwood, *Phys. Rev. Lett.* **80**, 4060 (1998).
34. J.R. Proot, C. Delerue, and G. Allan, *Appl. Phys. Lett.* **61**, 1948 (1992).
35. T. Takagahara and K. Takeda, *Phys. Rev. B* **46**, 15578 (1992).
36. M. S. Hybertsen, *Phys. Rev. Lett.* **72**, 1514 (1994).
37. Y. Kanemitsu, H. Uto, Y. Masumoto, T. Matsumoto, T. Futagi, and H. Mimura, *Phys. Rev. B* **48**, 2827 (1993).
38. S. Furukawa and T. Miyasato, *Phys. Rev. B* **38**, 5726 (1988).
39. C. Delerue, G. Allan, and M. Lannoo, *Phys. Rev. B* **48**, 11024 (1993).
40. M. V. Wolkin, J. Lorne, P. M. Fauchet, G.Allan, and C. Delerue, *Phys. Rev. Lett.* **82**, 197 (1999).
41. R. Boukherroub, S. Morin, D. D. M. Wayner, and D. J. Lockwood, *Phys. Stat. Sol. (a)*, **182**, 117 (2000).
42. R. Boukherroub, S. Morin, D. D. M. Wayner, F. Bensebba, G. I. Sproule, J.-M. Baribeau, and D. J. Lockwood, *Chem. Mater.* **13**, 2002 (2001).
43. R. Boukherroub, D. D. M. Wayner, D. J. Lockwood, and L. T. Canham, *J. Electrochem. Soc.* **148**, H91 (2001).
44. R. Boukherroub, J. T. C. Wojtyk, D. D. M. Wayner, and D. J. Lockwood, *J. Electrochem. Soc.* **149**, H59 (2002).
45. R. Boukherroub, D.D.M. Wayner and D.J. Lockwood, *Appl. Phys. Lett.* **81**, 601 (2002).
46. B. Gelloz, H. Sano, R. Boukherroub, D. D. M. Wayner, D. J. Lockwood, and N. Koshida, *Appl. Phys. Lett.* **83**, 2342 (2003).
47. R. Tsu, *Nature* **364**, 19 (1993).
48. C. J. Frosch and L. Derick, *J. Electrochem. Soc.* **104**, 547 (1957).
49. D. J. Lockwood, *Phase Transitions* **68**, 151 (1999).
50. Z. H. Lu, D. J. Lockwood, and J.-M. Baribeau, *Nature* **378**, 258 (1995).
51. D. J. Lockwood, Z. H. Lu, and J.-M. Baribeau, *Phys. Rev. Lett.* **76**, 539 (1996).
52. A. Zunger and L.-W. Wang: *Appl. Surf. Sci.* **102**, 350 (1996).
53. Z. H. Lu, D. J. Lockwood, and J.-M. Baribeau, *Solid-State Electron.* **40**, 197 (1996).
54. R. Tsu, A. Filios, C. Lofgren, K. Dovidenko, and C. G. Waugh, *Electrochem. Solid-State Lett.* **1**, 80 (1998).
55. D. J. Lockwood, J.-M. Baribeau, and Z. H. Lu, in *Advanced Luminescent Materials*, D. J. Lockwood, P. M. Fauchet, N. Koshida, and S. R. J. Brueck, Eds., p. 339, The Electrochemical Society Proceedings Series, Pennington, NJ (1996).

56. J. A. Brum and G. Bastard, *J. Phys. C. Solid State Phys.* **18**, L789 (1985).

57. Z. H. Lu, J.-M. Baribeau, D. J. Lockwood, M. Buchanan, N. Tit, C. Dharma-wardana, and G. C. Aers, *SPIE Proc.* **3491**, 457 (1998).

58. B. T. Sullivan, D. J. Lockwood, H. J. Labbé, and Z.-H. Lu, *Appl. Phys. Lett.* **69**, 3149 (1996).

59. D. J. Lockwood, B. T. Sullivan, and H. J. Labbé, *J. Lumin.* **80**, 75 (1999).

60. D. J. Lockwood and L. Tsybeskov, *J. Nanophotonics* **2**, 022501 (2008).

61. R. Tsu, Q. Zhang, and A. Filios, *SPIE Proc.* **3290**, 246 (1997).

62. A. G. Nassiopoulou, V. Ioannou-Sougleridis, P. Photopoulos, A. Travlos, V. Tsakiri, and D. Papadimitriou, *Phys. Stat. Sol. (a)* **165**, 79 (1998).

63. G. G. Qin, S. Y. Ma, Z. C. Ma, W. H. Zong, and L. P. You, *Solid State Commun.* **106**, 329 (1998).

64. L. Heikkilä, T. Kuusela, H.-P. Hedman, and H. Ihantola, *Appl. Surf. Sci.* **133**, 84 (1998).

65. V. Ioannou-Sougleridis, V. Tsakiri, A. G. Nassiopoulou, P. Photopoulos, F. Bassani, and F. Arnaud d'Avitaya, *Phys. Stat. Sol. (a)* **165**, 97 (1998).

66. M. Zacharias and P. Streitenberger, *Phys. Rev. B* **62**, 8391 (2000).

67. Y. Takahashi, T. Furuta, Y. Ohno, T. Ishiyama, and M. Tabe, *Jpn. J. Appl. Phys.* **34**, 950 (1995).

68. Y. Kanemitsu and S. Okamoto, *Phys. Rev. B* **56**, R15561 (1997).

69. E.-C. Cho, M. A. Green, R. Corkish, P. Reece, M. Gal, and S.-H. Lee, *J. Appl. Phys.* **101**, 063105 (2007).

70. Z. H. Lu and D. Grozea, *Appl. Phys. Lett.* **80**, 255 (2002).

71. N. Sato and T. Yonehara, *Appl. Phys. Lett.* **65**, 1924 (1994).

72. Z. H. Lu, J.-M. Baribeau and D. J. Lockwood, *J. Appl. Phys.* **76**, 3911 (1994).

73. D. J. Lockwood, Z. H. Lu, and D. Grozea, in *Advanced Luminescent Materials and Quantum Confinement II*, M. Cahay, J. P. Leburton, D. J. Lockwood, S. Bandyopadhyay, N. Koshida, and M. Zacharias, Eds., p. 159, The Electrochemical Society Proceedings Series, Pennington, NJ (2002).

74. D. J. Lockwood, Z. H. Lu and D. Grozea, *SPIE Proc.* **4808**, 40 (2002).

75. D. J. Lockwood, Z. H. Lu and D. Grozea, *Inst. Phys. Conf. Series* **171**, G3.1 (2003).

76. D. J. Lockwood, M. W. C. Dharma-wardana, Z. H. Lu, D. H. Grozea, P. Carrier and L. J. Lewis, *Mat. Res. Soc. Proc.* **737**, 243 (2003).

77. D. J. Lockwood, R. L. Williams and Z.-H. Lu, in *Nanoscale Devices, Materials, and Biological Systems: Fundamentals and Applications*, M. Cahay, M. Urquidi-Macdonald, S. Bandyopadhyay, P. Guo, H. Hasegawa, N. Koshida, J. P. Leburton, D. J. Lockwood, S. Seal and A. Stella, Editors, p. 243, The Electrochemical Society Proceedings Series, Pennington, NJ (2005).

78. B. K. Agrawal and S. Agrawal, *Appl. Phys. Lett.* **77**, 3039 (2000).

79. P. Carrier, L. J. Lewis, and M. W. C. Dharma-wardana, *Phys. Rev. B* **65**, 165339 (2002).

80. C. Delerue, G. Allan, and M. Lannoo, in *Light Emission in Silicon*, D. J. Lockwood, Ed., Chapter 7, Academic Press, New York (1998).

81. I. Vasiliev, S. Ogut, and J. R. Chelikowsky, *Phys. Rev. B* **65**, 115416 (2002).

82. S. Zollner, D. J. Lockwood, and Z. H. Lu (private communication).

83. D. J. Lockwood, in *Light Emission in Silicon*, D. J. Lockwood, Ed., Academic Press, New York (1998).

84. Y. Kanemitsu, in *Light Emission in Silicon*, D. J. Lockwood, Ed., Chapter 5, Academic Press, New York (1998).

85. S. Saito, D. Hisamoto, H. Shimizu, H. Hamamura, R. Tsuchiya, Y. Matsui, T. Mine, T. Arai, N. Sugii, K. Torii, S. Kimura, and T. Onai, *Appl. Phys. Lett.* **89**, 163504 (2006).

86. S. Saito, Y. Suwa, H. Arimoto, N. Sakuma, D. Hisamoto, H. Uchiyama, J. Yamamoto, T. Sakamizu, T. Mine, S. Kimura, T. Sugawara, and M. Aoki, *Appl. Phys. Lett.* **95**, 241101 (2009).

87. M. d'Avezac, J.-W. Luo, T. Chanier, and A. Zunger, *Phys. Rev. Lett.* **108**, 027401 (2012).

88. H. J. Xiang, B. Huang, E. Kan, S.-H. Wei, and X. G. Gong, *Phys. Rev. Lett.* **110**, 118702 (2013).

89. D. J. Lockwood, N. L. Rowell, L. Favre, A. Ronda and I. Berbezier, *ECS J. Solid State Sci. Tech.*, **7**(8), R115 (2018).

90. J. Weber and M. I. Alonso, *Phys. Rev. B* **40**, 5683 (1989).

91. R. Sauer, J. Weber, J. Stolz, E. R. Weber, K.-H. Kfisters, and H. Alexander, *Appl. Phys. A* **36**, 1 (1985).

92. W. M. Duncan, P.-H. Chang, B.-Y. Mao, and C.-E. Chen, *Appl. Phys. Lett.* **51**, 773 (1987).

93. G. Davies, *Phys. Reports* **176**, 83 (1989).

94. S. Fukatsu, Y. Mera, M. Inoue, K. Maeda, H. Akiyama, and H. Sakaki, *Appl. Phys. Lett.* **68**, 1889 (1996).

95. N. L. Rowell, J.-M. Baribeau, and D. C. Houghton, in *Proceedings of the Second International Symposium on Silicon Molecular Beam Epitaxy*, J. C. Bean and L. J. Schowalter, Eds., p. 48, *Electrochem. Soc. Proc.*, **88**(8), (1988).

96. D. J. Lockwood and L. Tsybeskov, in *Handbook of Silicon Photonics*, L. Vivien and L. Pavesi, Eds., pp. 354-371, CRC Press, Boca Raton, FL (2013).

97. N. Modi, D. J. Lockwood, J.-M. Baribeau, X. Wu, and L. Tsybeskov, *J. Appl. Phys.* **111**, 114313 (2012).

98. B. V. Kamenev, L. Tsybeskov, J.-M. Baribeau, and D. J. Lockwood, *Phys. Rev. B* **72**, 193306 (2005).

99. B. V. Kamenev, L. Tsybeskov, J.-M. Baribeau, and D. J. Lockwood, *Appl. Phys. Lett.*, **84**, 1293 (2004).

100. E.-K. Lee, D. J. Lockwood, J.-M. Baribeau, A. M. Bratkovsky, T. I. Kamins, and L. Tsybeskov, *Phys. Rev. B* **79**, 233307 (2009).

101. B. Ohnesorge, M. Albrecht, J. Oshinowo, A. Forchel, and Y. Arakawa, *Phys. Rev. B* **54**, 11532 (1996).

102. S. Fukatsu, Y. Mera, M. Inoue, K. Maeda, H. Akiyama, and H. Sakaki, *Appl. Phys. Lett.* **68** (1996).

103. A. Zrenner, B. Fröhlich, J. Brunner, and G. Abstreiter, *Phys. Rev. B* **52**, 16608 (1995).

104. I. Kuskovsky, G. F. Neumark, V. N. Bondarev, and P. V. Pikhitsa, *Phys. Rev. Lett.* **80**, 2413 (1998).

105. D. G. Thomas, J. J. Hopfield, and W. M. Augustyniak, *Phys. Rev.* **140**, A202 (1965).

106. J. Xiang, W. Lu, Y. Hu, Y. Wu, H. Yan, and C. M. Lieber, *Nature* **441**, 489 (2006).

107. G. Conibeer, M. Green, R. Corkish, Y. Cho, E. C. Cho, C. W. Jiang, T. Fangsuwannarak, E. Pink, Y. D. Huang, T. Puzzer, T. Trupke, B. Richards, A. Shalav, and K. L. Lin, *Thin Solid Films* **511**, 654 (2006).

108. P. Werner, N. D. Zakharova, G. Gertha, L. Schuberta, and U. Gösele, *Int. J. Mat. Res.* **97**(7), 1008 (2006).
109. E. Sutter, B. Ozturk, and P. Sutter, *Nanotechnology* **19**, 435607 (2008).
110. J. P. Noel, N. L. Rowell, and J. E. Greene, *J. Appl. Phys.* **77**, 4623 (1995).
111. Y. S. Tang, C. D. Wilkinson, C. M. Sotomayor Torres, D. W. Smith, T. E. Whall, and E. H. C. Parker, *Solid State Commun.* **85**, 199 (1993).
112. E. G. Barbagiovanni, D. J. Lockwood, P. J. Simpson, and L. V. Goncharova, *Appl. Phys. Rev.* **1**(1), 011302 (2014).
113. N. L. Rowell, J.-P. Noël, D. C. Houghton, A. Wang, L. C. Lenchyshyn, M. L. W. Thewalt, and D. D. Perovic, *J. Appl. Phys.*, **74**, 2790 (1993).
114. N. L. Rowell, D. J. Lockwood, D. C. Houghton, J.-P. Noël, and J.-M. Baribeau, *ECS J. Solid State Sci. Tech.* **7**(12), R195 (2018).
115. J. P. Dismukes, L. Ekstrom and R. J. Paff, *J. Phys. Chem.* **68**, 3021 (1964).
116. K. Guilloy, N. Pauc, A. Gassenq, Y. M. Niquet, J. M. Escalante, I. Duchemin, S. Tardif, G. Osvaldo Dias, D. Rouchon, J. Widiez, J. M. Hartmann, R. Geiger, T. Zabel, H. Sigg, J. Faist, A. Chelnokov, V. Reboud, and V. Calvo, *ACS Photonics* **3**, 1907 (2016).
117. H. K. Shin, D. J. Lockwood, and J. M. Baribeau, *Solid State Commun.* **114**, 505 (2000).
118. H. Chen, Y. K. Li, C. S. Peng, H. F. Liu, Y. L. Liu, Q. Huang, J. M. Zhou, and Qi-Kun Xue, *Phys. Rev. B* **65**, 233303 (2002).
119. M. Grydlik, F. Hackl, H. Groiss, M. Glaser, A. Halilovic, T. Fromherz, W. Jantsch, F. Schäffler, and M. Brehm, *ACS Photonics* **3**, 298 (2016).
120. M. Grydlik, M. T. Lusk, F. Hackl, A. Polimeni, T. Fromherz, W. Jantsch, F. Schäffler, and M. Brehm, *Nano Lett.* **16**, 6802 (2016).
121. P. Rauter, L. Spindlberger, F. Schäffler, T. Fromherz, J. Freund, and M. Brehm, *ACS Photonics* **5**(2), 431 (2017).
122. D. J. Lockwood, *ECS Trans.* **50**(37), 39 (2013).
123. V. Rutckaia, F. Heyroth, A. Novikov, M. Shaleev, M. Petrov, and J. Schilling, *Nano Lett.* **17**, 6886 (2017).
124. T. A. Amollo, G. T. Mola, and V. O. Nyamori, *Nanotechnology* **28**, 495703 (2017).
125. L. E. Vorobjev, D. A. Firsov, V. Yu Panevin, A. N. Sofronov, R. M. Balagula, and A. A. Tonkikh, *J. Phys: Conf. Series* **586**, 012001 (2015).
126. E. M. F. Vieira, J. Toudert, A. G. Rolo, A. Parisini, J. P. Leitão, M. R. Correia, N. Franco, E. Alves, A. Chahboun, J. Martín-Sánchez, R. Serna, and M. J. M. Gomes, *Nanotechnology* **28**, 345701 (2017).
127. M. Greben, P. Khoroshyy, S. Gutsch, D. Hiller, M. Zacharias, and J. Valenta, *Beilstein J. Nanotechnol.* **2017**(8), 2315 (2017).
128. J.-W. Luo, S.-S. Li, I. Sychugov, F. Pevere, J. Linnros, and A. Zunger, *Nature Nanotech.* **12**, 930 (2017).
129. W. de Boer, D. Timmerman, I. Yassievich, A. Capretti, and T. Gregorkiewicz, *Nature Nanotech.* **12**, 932 (2017).
130. W. D. A. M. de Boer, D. Timmerman, K. Dohnalová, I. N. Yassievich, H. Zhang, W. J. Buma, and T. Gregorkiewicz, *Nature Nanotech.* **5**, 878 (2010).
131. D. Kovalev, *Nature Nanotech.* **5**, 827 (2010).

Chapter 2

Materials and Devices for Novel Non-Volatile Memory Elements and Neuromorphic Computing

ECS Transactions, 89 (3) 39-44 (2019)
10.1149/08903.0039ecst ©The Electrochemical Society

Multilevel Resistive Switching in Hf-based RRAM

B. Jain,[1] C.S. Huang,[1] D. Misra,[1] K. Tapily,[2] R.D. Clark,[2] S. Consiglio,[2] C.S. Wajda,[2] and G.J. Leusink[2]

[1]ECE Dept, NJIT, Newark, NJ 07102, USA
[2]TEL Technology Center America, LLC, Albany, NY 12203, USA

In this paper, the multilevel switching behaviors of resistive random-access memory (RRAM) devices with three different dielectric materials such as HfO_2, $HfZrO_2$ and $HfAlO_2$ are investigated. We have further explored the switching characteristics with two different top electrode materials and with different processing environments. In all devices we have introduced a thin buffer layer to reduce switching power and improve the uniformity. Variation in the resistive behavior (Roff/Ron values) of different RRAM structures were observed and was correlated with possible oxygen vacancy related defects present in the dielectric.

Introduction

As the conventional flash memory technology is facing several challenges like scalability issues, slow programming speed and high-power consumption, various alternative technologies for future non-volatile memories are being investigated (1). Resistive random-access memories (RRAM) are one of the most promising candidates owing to its simple structure, low power consumption, high density, fast switching speed, excellent scalability, high endurance, good retention and compatible with the CMOS technology (2). The structure of RRAM device consists of a dielectric layer sandwiched between a top and a bottom electrode. The resistance of RRAM can be reversibly switched in between high and low resistance states by the application of an external voltage. Therefore, the dielectric layer plays an important role for the RRAM switching. Recently, many transition metal oxide (TMO) based RRAM devices are reported to exhibit nonvolatile data storage potential (3-9). HfO_2 is extensively studied for the RRAM application. However, stoichiometric HfO_2 is not suitable because of lack of enough oxygen vacancies and requires high forming voltage that leads to higher power consumption (10). Therefore, different HfO_2 based dielectric materials are investigated to improve the performance of RRAM devices (11-14). Dielectrics like $HfZrO_2$ and $HfAlO_2$ needs to be studied for advanced RRAM devices that can have a required switching behavior and possible multilevel storage for high density data storage application.

In this work, we investigated the multi-level storage characteristics of $HfZrO_2$ and $HfAlO_2$ in RRAM devices and compared with HfO_2. We have also explored the switching characteristics by varying top electrode materials in some devices and varying processing conditions. The measurements were performed by using a dc voltage sweeping mode for different compliance currents. Since variation in the oxygen vacancies distribution in the dielectric region under the application of electric field is responsible for the filament formation, we have correlated the switching characteristics with the oxygen vacancy related defects present in the dielectric layer.

Experimental

To enable a high device reliability and a deeper understanding of the material parameters controlling the switching characteristics, the dielectrics were compared at fixed geometry and top/bottom electrode types. For device fabrication, 7nm of HfO_2, 7nm of $HfZrO_2$ and 7nm of $HfAlO_2$ were deposited by atomic layer deposition (ALD) on the top of 10nm-Ti/50nm-TiN/1nm-Al_2O_3 layer. ALD $Hf_{1-x}Zr_xO_2$ is deposited in a 300 mm TEL Trias cleanroom tool by using tetrakis (ethylmethylamido) hafnium as the Hf precursor, tetrakis (ethylmethylamido) zirconium as the Zr precursor and H_2O as the oxidant at a deposition temperature of 250˚C. The $Hf_{1-x}Zr_xO_2$ films were deposited by precisely controlling the individual HfO_2 and ZrO_2 ALD cycles contained within each super cycle of the ALD process. The Hf-precursor to Zr-precursor pulse ratio of 1:1 was used for $Hf_{1-x}Zr_xO_2$ whereas samples with only HfO_2 were deposited without any Zr-precursor. Details of the fabrication process can be found elsewhere (15,16). ALD $HfAlO_x$ was deposited by using tetrakis (ethylmethylamino) hafnium (TEMAH) and trimethylaluminum (TMA) as Hf and Al precursors with H_2O as the co-reactant at 250^0C deposition temperature. Thereafter, 8nm-Ti/6nm-ALD-TiN plus 50nm PVD-TiN layer were deposited on top of the dielectric layer. Insets in Fig. 1(a)-(c) shows the device structures. To further understand the effect of processing variation, RRAM devices with $HfZrO_2$ as dielectric material was subjected to post deposition annealing (PDA) at 700˚C for 60sec in the presence of nitrogen. The role of top electrode was investigated by adding a 2nm ALD TiN layer prior to 8nm Ti layer in device with $HfZrO_2$. In all our RRAM structures, a thin Al_2O_3 (1 nm) buffer layer is deposited prior to depositing the switching layer. It has been found that Al_2O_3 have shown to reduce the programming current and enhanced switching uniformity and stability in the switching behavior when embedded between the switching layer and the bottom electrode (17-19).

Results and Discussion

Fig. 1(a)-(c) shows the I-V characteristics of the three RRAM structures with the $HfAlO_2$ based RRAM (a), $HfZrO_2$ based RRAM (b) and HfO_2 based RRAM (c) as a function of increasing compliance currents. $HfAlO_2$ and $HfZrO_2$ based RRAM devices show increase in the low resistance state (LRS) current levels (I_{LRS}) when the compliance current (CC) was increased as shown in Fig. 1(a)-(b), clearly showing a decrease in resistance. Whereas, no multilevel LRS current states with increase in CC was observed in HfO_2 based RRAM devices as shown in Fig. 1(c). The reason for poor switching characteristics in HfO_2 based devices is possibly due to lack of oxygen vacancies available in HfO_2. It is known that in stoichiometric HfO_2, with reduced number of available oxygen vacancies (10,14), formation of a thicker filament at a higher CC is difficult to achieve. Our HfO_2 based RRAM samples constitutes highly stoichiometry HfO_2. We believe, low oxygen vacancy concentration limits a thicker filament formation at higher CC for these devices. In switching dielectrics, higher oxygen vacancy concentration is an important means to improve the performance of the RRAM device (20-22). When HfO_2 was alloyed with Al or Zr, an improvement in the formation of thicker conducting filament was observed (decrease in resistance) due to increase in the number of defects sites or oxygen vacancies.

To reaffirm the contribution of oxygen vacancies in case of $HfZrO_2$ and $HfAlO_2$ based RRAM devices, I-V characteristics of intrinsic devices were plotted for all the three RRAM devices in the pristine high resistive states as shown in the Fig. 2. The initial current is normally due to the existing traps, originating from oxygen vacancies, present in the switching dielectric. As the current in case of $HfZrO_2$ based RRAM is higher as compared to other two devices, it can be confirmed that despite the chemical equivalency of Hf and Zr, an apparent partial substitution of Hf by Zr leads to higher oxygen vacancy.

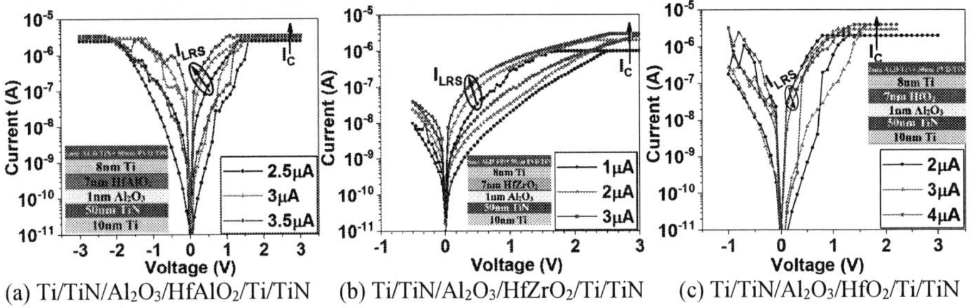

(a) Ti/TiN/Al₂O₃/HfAlO₂/Ti/TiN   (b) Ti/TiN/Al₂O₃/HfZrO₂/Ti/TiN   (c) Ti/TiN/Al₂O₃/HfO₂/Ti/TiN

Figure 1. I-V characteristics (a)-(c) of HRS/LRS resistance of Ti/TiN/Al₂O₃/HfAlO₂/Ti/TiN, Ti/TiN/Al₂O₃/HfZrO₂/Ti/TiN and Ti/TiN/Al₂O₃/HfO₂/Ti/TiN devices for increasing compliances respectively.

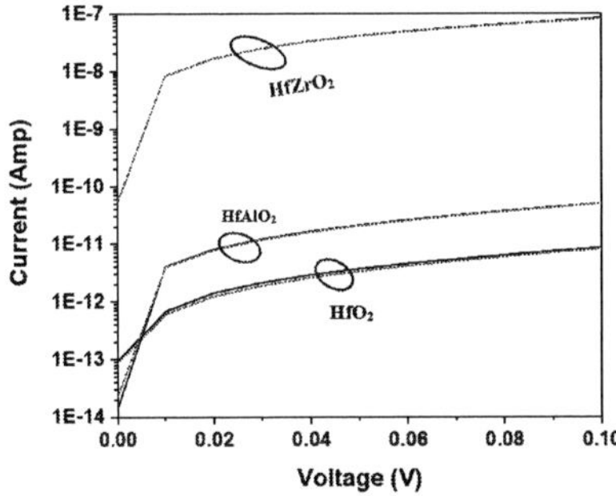

Figure 2. I-V characteristics of $HfZrO_2$, $HfAlO_2$ and HfO_2 based RRAM in pristine states.

Fig. 3(a)-(c) shows cumulative distribution of DC endurance cycling on the LRS/HRS resistances of the three RRAM structures at different compliance currents. The endurance cycling was performed with 20 cycles per compliance level (60 cycles, in total). As can be seen from the figure, different LRS levels are well resolved after dc endurance cycling for $HfAlO_2$ and $HfZrO_2$ based RRAM devices. Although multilevel switching is observed in both $HfAlO_2$ and $HfZrO_2$ based RRAM, it is more prominent in the $HfZrO_2$ RRAM case.

$HfZrO_2$ based RRAM is further analyzed to understand the effect of post deposition annealing (PDA) on the multilevel switching of RRAM devices. Fig. 4 shows the cycle to cycle cumulative distribution of HRS/LRS resistance of $HfZrO_2$ based RRAM with PDA, Fig. 4(a) and without PDA, Fig. 4(b) with increasing compliance currents. It can be seen from Fig. 4 that after PDA, distribution spread of the HRS, R_{off} and LRS R_{on} resistances is reduced i.e. cycle to cycle variability is improved at the same CC level. Also, the R_{on}/R_{off} values are more consistent with the number of cycles. It is known that when HfZrO$_2$ is subjected to PDA, film densification and thickness reduction occur [23, 26]. Due to the thickness reduction of the dielectric, more oxygen ions were attracted towards the Ti top electrode which created additional oxygen vacancies inside the dielectric region [23-25]. However, the R_{on} values of low compliance current was identical to that of R_{off} values of high compliance current making it difficult to distinguish the multilevel states. A reasonably large resistance window is essential in RRAM to allow for enough read margin between the HRS and the LRS, ensuring enough immunity from post programming fluctuations of resistance.

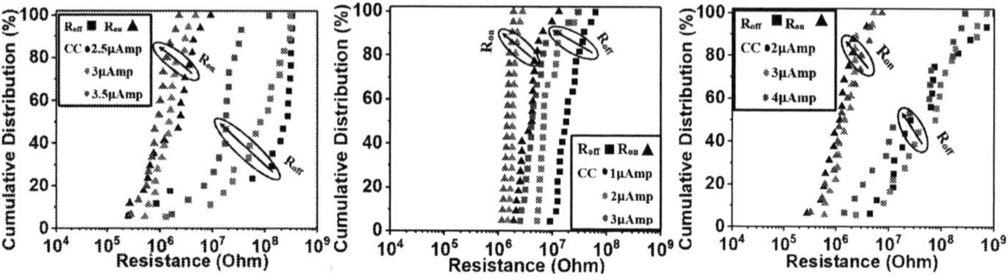

(a) Ti/TiN/Al$_2$O$_3$/HfAlO$_2$/Ti/TiN   (b) Ti/TiN/Al$_2$O$_3$/HfZrO$_2$/Ti/TiN   (c) Ti/TiN/Al$_2$O$_3$/HfO$_2$/Ti/TiN

Figure 3. Cycle-to-cycle distributions of HRS/LRS resistance (a)-(c) of Ti/TiN/Al$_2$O$_3$/HfAlO$_2$/Ti/TiN, Ti/TiN/Al$_2$O$_3$/HfZrO$_2$/Ti/TiN and Ti/TiN/Al$_2$O$_3$/HfO$_2$/Ti/TiN devices for increasing compliances respectively.

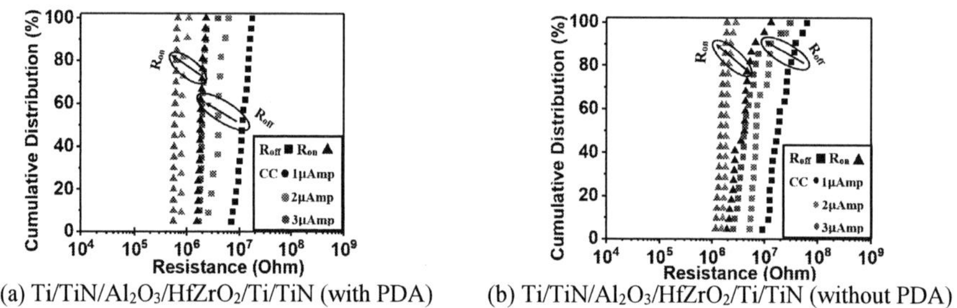

(a) Ti/TiN/Al$_2$O$_3$/HfZrO$_2$/Ti/TiN (with PDA)   (b) Ti/TiN/Al$_2$O$_3$/HfZrO$_2$/Ti/TiN (without PDA)

Figure 4. The cycle-to-cycle distributions of HRS/LRS resistance of (a) Ti/TiN/Al$_2$O$_3$/HfZrO$_2$/Ti/TiN (with PDA), (b) Ti/TiN/Al$_2$O$_3$/HfZrO$_2$/Ti/TiN (without PDA) devices for increasing compliances.

To improve the resistance window and to decrease the cycle variability the top electrode of HfZrO$_2$ based RRAM was modified by adding a 2nm-thin TiN layer prior to Ti layer deposition. Fig. 5 shows the cycle to cycle distribution of HRS/LRS resistance of HfZrO$_2$ based RRAM without (Fig. 5(a)) and with (Fig. 5 (b)) additional TiN layer at different compliance currents. Effect of adding an additional TiN layer to the HfZrO$_2$ based RRAM is clearly visible in the Fig. 5 where the Ron and Roff values are distinctly separated. The possible reason for the observed electrode material dependence of resistive switching behavior is due to the work function difference of the electrode material (28, 29). The work function difference between the two top metal electrodes and the interface layers can be attributed to the observed variation in switching characteristics.

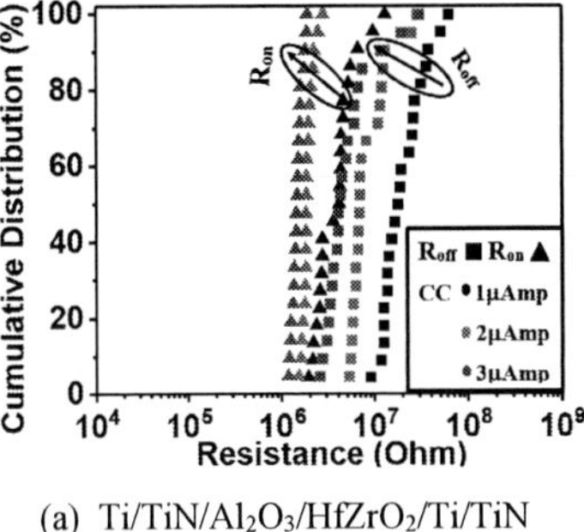
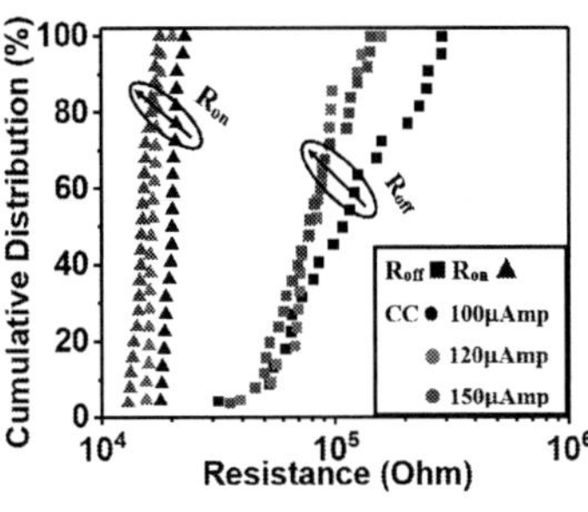

(a) Ti/TiN/Al$_2$O$_3$/HfZrO$_2$/Ti/TiN   (b) Ti/TiN/Al$_2$O$_3$/HfZrO$_2$/TiN/Ti/TiN

Figure 5. The cycle-to-cycle distributions of HRS/LRS resistance of (a) Ti/TiN/Al$_2$O$_3$/HfZrO$_2$/Ti/TiN, (b) Ti/TiN/Al$_2$O$_3$/HfZrO$_2$/TiN/Ti/TiN devices for increasing compliances.

The work function of TiN (~ 4.5 eV) and Ti (~ 4.3 eV) and electron affinities of TiO$_2$ (~3.2eV) (27) and HfZrO$_2$ (~2.7eV) needs to be carefully evaluated. Oxidized Ti and TiN layers are believed to be formed when a positive voltage is applied on top electrode due to the attraction of oxygen ions towards the top electrode. As Ti is more reactive than TiN, effect of oxidized TiN layer is insignificant. This increases the oxygen vacancy concentration in Ti/HfZrO$_2$ as compared to TiN/HfZrO$_2$ based RRAM structure due to high reactivity of Ti metal. This leads to the overlapping of high and low resistance values. Whereas, in case of TiN as the oxygen vacancy concentration being less at the interface, the formation of filament is more controlled. This is further supported by increased in compliance current for switching in TiN/HfZrO$_2$ based RRAM structures.

In summary, the multilevel switching behavior of RRAM devices with different dielectric materials, HfZrO$_2$ and HfAlO$_2$, were studied and compared with HfO$_2$. It was observed that intrinsic concentration of oxygen vacancies dominates the switching characteristics. As compared to HfO$_2$ and HfAlO$_2$, multilevel switching was found to be more prominent in case of HfZrO$_2$ based RRAM devices. Furthermore, it was found that, the multilevel switching characteristics of HfZrO$_2$ based RRAM was improved when the dielectrics was subjected to post deposition annealing in the presence of N$_2$. Use of TiN instead of Ti as the top electrode materials enhances the multilevel storage capabilities due to the work function difference leading to oxidation at the interface

References

1. D. Ielmini, *Semiconductor Science and Technology* **31**, 063002 (2016).
2. E. Ambrosi, A. Bricalli, M. Laudato, and D. Ielmini, *Faraday Discussions* (2018).
3. A. Prakash and H. Hwang, *Physical Sciences Reviews* **1**, (2016).
4. F. Zhou, Y.-F. Chang, B. Fowler, K. Byun, and J.C. Lee, *Applied Physics Letters* **106**, 063508 (2015).

5. B. Chakrabarti, R.V. Galatage, and E.M. Vogel, *IEEE Electron Device Letters* **34**, 867 (2013).
6. Y. Li, S. Long, Q. Liu, Q. Wang, M. Zhang, H. Lv, L. Shao, Y. Wang, S. Zhang, Q. Zuo, S. Liu, and M. Liu, *Physica Status Solidi (RRL) - Rapid Research Letters* **4**, 124 (2010).
7. S.-Y. Wang, C.-W. Huang, D.-Y. Lee, T.-Y. Tseng, and T.-C. Chang, *Journal of Applied Physics* **108**, 114110 (2010).
8. W. Chen, W. Lu, B. Long, Y. Li, D. Gilmer, G. Bersuker, S. Bhunia, and R. Jha, *Semiconductor Science and Technology* **30**, 075002 (2015).
9. S.K. Vishwanath, H. Woo, and S. Jeon, *Nanotechnology* **29**, 235202 (2018).
10. D.C. Gilmer, G. Bersuker, H.-Y. Park, C. Park, B. Butcher, W. Wang, P.D. Kirsch, and R. Jammy, *2011 3rd IEEE International Memory Workshop (IMW)* (2011).
11. Z. Hou, Z. Wu, and H. Yin, *ECS Journal of Solid-State Science and Technology* **7**, (2018).
12. P. Zhou, L. Ye, Q.-Q. Sun, L. Chen, S.-J. Ding, A.-Q. Jiang, and D.W. Zhang, *IEEE Transactions on Nanotechnology* **11**, 1059 (2012).
13. L. Chen, Y. Xu, Q.-Q. Sun, P. Zhou, P.-F. Wang, S.-J. Ding, and D.W. Zhang, *IEEE Electron Device Letters* (2010).
14. D. Misra, S. Sultana, B. Jain, N. Bhat, K. Tapily, R.D. Clark, S. Consiglio, C.S. Wajda, and G.J. Leusink, *ECS Transactions* **86**, 77 (2018).
15. M. Bhuyian, D. Misra, K. Tapily, R. Clark, S. Consiglio, C. Wajda, G. Nakamura, and G. Leusink, *ECS Transactions* **61**, 41 (2014).
16. K. Tapily, S. Consiglio, R.D. Clark, R. Vasić, C.S. Wajda, J. Jordan-Sweet, G.J. Leusink, and A.C. Diebold, *ECS Journal of Solid State Science and Technology* **4**, (2014).
17. S. Yu, Y. Wu, Y. Chai, J. Provine, and H.-S.P. Wong, *Proceedings of 2011 International Symposium on VLSI Technology, Systems and Applications* (2011).
18. S. Stathopoulos, A. Khiat, M. Trapatseli, S. Cortese, A. Serb, I. Valov, and T. Prodromakis, *Scientific Reports* **7**, (2017).
19. S. Yu, B. Gao, H. Dai, B. Sun, L. Liu, X. Liu, R. Han, J. Kang, and B. Yu, *Electrochemical and Solid-State Letters* **13**, (2010).
20. Y.S. Chen, B. Chen, B. Gao, L.F. Liu, X.Y. Liu, and J.F. Kang, *Journal of Applied Physics* **113**, 164507 (2013).
21. W. Lian, S. Long, H. Lü, Q. Liu, Y. Li, S. Zhang, Y. Wang, Z. Huo, Y. Dai, J. Chen, and M. Liu, *Chinese Science Bulletin* **56**, 461 (2011).
22. B. Gao, H.W. Zhang, S. Yu, B. Sun, L.F. Liu, X.Y. Liu, Y. Wang, R.Q. Han, J.F. Kang, B. Yu, Y.Y. Wang, *In VLSI Technology, 2009 Symposium on*, 30 (2009)
23. K. Nakajima, A. Fujiyoshi, Z. Ming, M. Suzuki, and K. Kimura, *Journal of Applied Physics* **102**, 064507 (2007).
24. H. Shima and H. Akinaga, *2012 12th IEEE International Conference on Nanotechnology (IEEE-NANO)* (2012).
25. P.-S. Chen, H.-Y. Lee, Y.-S. Chen, F. Chen, and M.-J. Tsai, *Japanese Journal of Applied Physics* **49**, (2010).
26. M. N. Bhuyian, D. Misra, K. Tapily, R. D. Clark, S. Consiglio, C. S. Wajda, G. Nakamura, and G. J. Leusink, *ECS Journal of Solid-State Science and Technology*, **3(5)**, N83, 2014.
27. Y. Qi, C.Z. Zhao, C. Liu, Y. Fang, J. He, T. Luo, L. Yang, and C. Zhao, *Semiconductor Science and Technology* **33**, 045003 (2018).
28. H. Shima, M. Takahashi, Y. Naitoh, and H. Akinaga, *IEEE Journal of the Electron Devices Society* **1** (2018).
29. G.W. Dietz, W. Antpöhler, M. Klee, and R. Waser, *Journal of Applied Physics* **78**, 6113 (1995).

ECS Transactions, 89 (3) 45-51 (2019)
10.1149/08903.0045ecst ©The Electrochemical Society

Selectorless Oxide-based Resistive Switching Memory with Nonuniform Dielectric for Low Power Crossbar Array Applications

Ying-Chen Chen[a], Chao-Cheng Lin[b], Sungjun Kim[c], and Jack C. Lee[a]

[a]Department of Electrical and Computer Engineering, The University of Texas at Austin, Austin, Texas 78758, USA
[b]National Nano Device Laboratories, NAR Labs, Hsinchu, Taiwan
[c]School of Electronics Engineering, Chungbuk National University, Cheongju, 28644, Republic of Korea

Selectorless resistive random-access memory (RRAM) with self-rectifying behavior is presented as a way to suppress sneak path current without an additional transistor or switch device integration, which is beneficial for reducing the cost and complexity of crossbar array fabrication. In this work, selectorless 1R-only graphite-based memristor devices have been demonstrated by utilizing the nonlinear (NL) resistive switching (RS) characteristics with well-designed operation conditions. The early stage of cycling observation as seasoning effect is investigated, while the wear-out mechanism is also discussed for the selectorless 1R-only RRAM crossbar array for high storage class memory applications.

Introduction

The memristors in binary metal oxide structures have attracted attentions since the current nonvolatile memory (NVM) has been approaching the scaling limit. Memristors having excellent scalability, high speed, and low power operation are desired for current hardware needs and future artificial intelligence in-memory computational hardware platform. The resistive random-access memory (RRAM) utilizing various metal oxides (i.e., SiO_x (1-6), HfO_2 (7-9), CeO_x (10-14), and SiN_x (15-18)) attracted much attention since the current nonvolatile memory (NVM) e.g. flash memory has been approaching the scaling limit (19-24). In order to realize the high storage class memory, the three-dimensional crossbar array configurations have been proposed. However, the sneak path current (SPC) is a main problem in crossbar memory configuration. The sneak path current significantly affects the read/write operation, because each word line is connected in perpendicular directions to the word line where the interference current from neighbour cells cannot be avoided in the crossbar configurations. In order to suppress the sneak path current in the array application, the transistor, diode, or selector device is designed to be integrated next to the memory cell for reducing the SPC. In other words, the selector devices are essential to address the sneak path current issue in the one transistor-one resistor (1T-1R) or one selector-one resistor (1S-1R) architecture (Fig. 1) (25-29). However, this 1S1R configuration will significantly bring up the process complexity, cost and deteriorate the scalability. In this work, selectorless 1R-only RRAM has been demonstrated by utilizing the nonlinear (NL) resistive switching (RS) characteristics. The self-rectifying nature is able to prevent the sneak path current issue since the on-state is achieved only at the range of high voltage (but before RESET happened), and the abrupt decrease of the conductance at low read voltage region, which effectively suppress the inevitable reading error of a selected cell (SC) in V/3 schemes. This work not only presented a 1R-only graphite-based selectorless RRAM in additional device-level design freedom in NL, but also provided a mechanism understanding of reliability related early seasoning cycles for low power and high-density RRAM array applications.

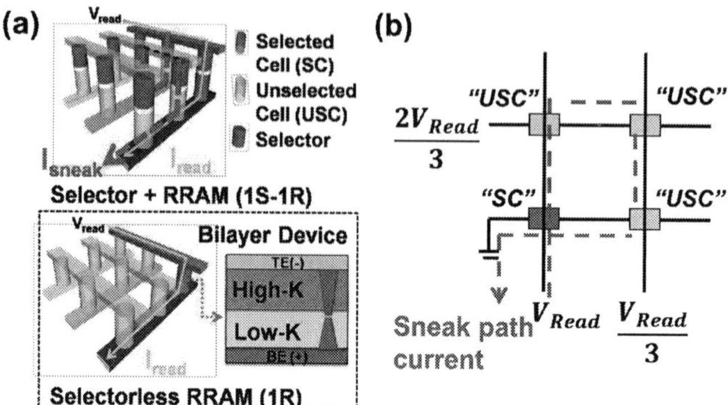

Fig. 1. Schematics of (a) 1S-1R configuration (i.e. 1 selector + 1 RRAM) and 1R-only selectorless RRAM with bilayer stacked structure. (b) sneak path current issue (red path) in the crossbar array application (V/3 read scheme).

Experiments

Fig. 2 (a) shows the tilt view and cross section view of scanning electron microscope image (SEM) of bilayer structure (H7G5 in the red dashed square). Selector-less memristors with varying side lengths of 400 nm, 600 nm, 800 nm, 1 μm have been fabricated. The starting substrates are heavily-doped N+ Si wafers. First, buffered-oxide-etch (BOE) was used to remove any native oxide layer on the substrate, followed by acetone and isopropyl alcohol (IPA) rinse. Then, titanium nitride of 200 nm thickness was deposited on N+ Si substrate as bottom electrode at a pressure of less than 10^{-5} Torr. Then, 5 nm of graphite and 7 nm of HfO_x were deposited as RS dielectric layers for bilayer structures (i.e. H7G5) by radio frequency (RF) sputtering. Pt (165 nm) was deposited as top electrodes. Meanwhile, memristors with a single HfO_x layer of 11 nm (i.e. H11) have been fabricated as reference devices. Conventional photolithography was used for patterning the top electrodes of various sizes using the lift-off method. Finally, Pt (165 nm)/HfO_x (7 nm)/ graphite (5 nm)/ TiN (200 nm) and Pt (165 nm)/HfO_x (11 nm)/ TiN (200 nm) devices were fabricated as bilayer and single devices. An Agilent B1500 and Lakeshore probe station were used for electrical characterization of all the graphite-based stacking RRAM devices.

Results and Discussion

The nonlinear characteristic observed in single RRAM devices can be used to solve the problem of interference from neighbor cells. The graphite-stacked RRAM has been confirmed with higher nonlinearity than a HfO_x single layer structure and proposed as a solution for sneak path issues in crossbar array applications. It depicts that using graphite thin film as a "low-k" stacking layer improves the nonlinearity characteristics. The nonlinearity is defined as the ratio of the current at fully read voltage (e.g. -0.6 V) to the current at 1/3 read voltage (e.g. -0.2 V). In other words, the selectorless RRAM with higher NL has a higher capability to avoid the sneak path current interferences. Voltage was applied to the bottom electrode (TiN) with the top electrode (Pt) at the ground. All the electrical testing was done in air environment instead of under vacuum as in previous works (30-34), which ruled out the ambient effect on the oxide-based devices. To initiate the RS in these devices, a one-step single sweep electroforming process was used (green curve in Fig. 2 (b)), i.e. a current-limited voltage-sweep to induce soft breakdown and a forward voltage sweep to electroform the device. In this work, the soft-breakdown process is performed by sweeping the voltage until current dramatically increases (i.e. ~4 V) to a compliance current limit (CCL) of 1 mA. Generally, this current compliance is required to prevent irreversible hard-breakdown of oxide layers during the forming process and increases the "electroforming yield" for RRAM devices, i.e. % of the good devices after electroforming. After the electroformation, RS performance is stabilized by cycling multiple times using DC voltage double sweeps. These stabilization cycles are called "seasoning cycles" in this work, which will be discussed in later sections. The SET process i.e. the switching from a high resistance state (HRS) to low-resistance state (LRS) takes place in positive polarity, while the RESET occurs in negative polarity. In this work, the SET voltage is ~1 V forward/reverse sweep with 1 mA CCL to program the device to LRS. The RESET process is stopped by sweeping voltage to -2.1 V for all structures (Fig. 1 (b)).

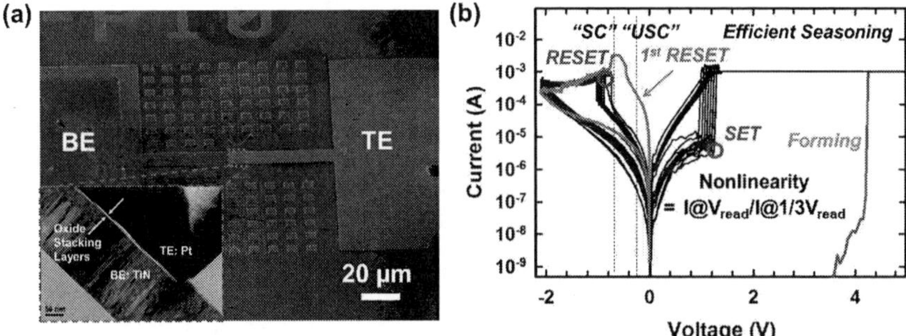

Fig. 2. (a) SEM images of bilayer selectorless RRAM devices. (b) I-V curve of the forming process (green curve), first RESET process (red curve) and switching cycles (black curves), for the efficient seasoning effect.

Figure 3 (a) shows the nonlinearity for 20 DC cycles (top panel) of the H7G5 device. The efficient seasoning effect shows a nonlinearity of >40 after the first 5 cycles after electroforming. The current at V_{read} is indicated as the blue curve, and the current at 1/3 V_{read} is indicated as the red curve. The reduction of current at 1/3 V_{read} results in higher nonlinearity i.e. efficient seasoning effect, which is not shown in the device with inefficient seasoning effect. The 1st RESET process shows a nonlinear increasing region (i.e. at -0.5V) in efficient seasoning effect of device H7G5. The nonlinear increasing region is not observed in the inefficient seasoning effect, in which the typical linear resistive switching in first RESET is observed. The current transport in this voltage range is found to fit well to the Fowler-Nordheim tunneling formula (red region in Fig. 3b and zoom-in plot (see inset) with scientifically accepted accuracy (R2 =99 %), which shows that less overshoot damage of forming may induce Fowler–Nordheim tunneling in the 1st RESET, as compared to devices with inefficient seasoning. The results also confirm that the seasoning effect determination with NL improvement by a robust low-k layer quality can improve the nonlinearity and selectorless device reliability.

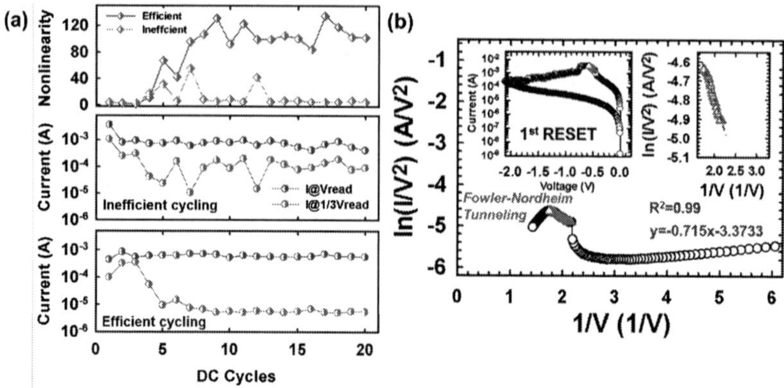

Fig. 3. (a) Nonlinearity with inefficient seasoning cycles and efficient seasoning cycles, after electroforming and the 1st RESET process. (b) Current transport mechanism understanding of 1ˢᵗ RESET process for the efficient seasoning.

The nonlinearity reliability characteristics in I_{LRS} at room temperature of H11 and H7G5 under SET CCL of 1 mA as shown in Fig. 4. With a thin graphite layer (5 nm) on the bottom of HfO_x (7 nm) layer, the NL is ~2x that of H4S9, and ~10x that of the single-layer HfO_x layer devices. The excellent sub-µs transitions with at least 10x resistance ratio after 1000 cycles at room temperature without any significant current change in selectorless RRAM devices are also observed (data not shown). This built-in nonlinearity makes graphite-stacked bilayer devices a potential candidate for 1R selectorless memristors. In addition, the NL of graphite-stacked memristors are optimized by adjusting SET CCL of 1 mA, as called "gap design method" for selectorless RRAMs (35-38).

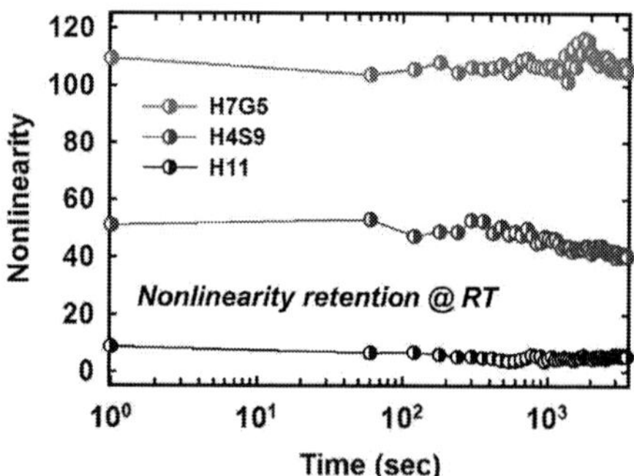

Fig. 4. Nonlinearity retention at room temperature HfO_x (7 nm)/Graphite (5 nm), HfO_x (4 nm)/ SiO_x (9 nm), and HfO_x (11 nm).

Conclusion

In this study, built-in NL characteristics have been realized for non-uniform dielectric selectorless memristors without a diode/selector. The relation between the first RESET with the following cycles is observed after the electroforming process. The Fowler-Nordheim tunneling mechanism is shown in the 1st RESET I-V curve, which results in a good nonlinearity (NL > 40) for the following switching cycles, i.e. efficient seasoning. To optimize the performance and reliability of self-rectifying behavior, an intrinsically design of layered functionality with mechanism insights are presented. The non-uniform dielectric memristors are promising for high-density, low-power selectorless RRAM array applications with good reliability properties. The highly NL characteristics observed in non-uniform dielectric bilayer devices with efficient seasoning process are desirable in suppressing sneak path current in crossbar arrays applications.

References

1. Y. F. Chang, B. Fowler, Y. C. Chen, Y. T. Chen, Y. Wang, F. Xue, F. Zhou, J. C. Lee, *J. Appl. Phys.* **116**, 043709 (2014). (DOI: 10.1063/1.4891244)
2. F. Zhou, Y. F. Chang, Y. Wang, Y. T. Chen, F. Xue, B. W. Fowler, J. C. Lee *Appl. Phys. Lett.* **105**, 163506 (2014). (DOI: 10.1063/1.4900422)
3. Y. F. Chang, B. Fowler, F. Zhou, K. Byun, J. C. Lee, *International Symposium on VLSI Technology, Systems, and Applications (VLSI-TSA)*, 1 (2015). (DOI: 10.1109/VLSI-TSA.2015.7117558)
4. Y. F. Chang, B. Fowler, F. Zhou, J. C. Lee, *2015 International Symposium on Next-Generation Electronics*, 1 (2015). (DOI: 10.1109/ISNE.2015.7131996).
5. Y. F. Chang, B. Fowler, Y. C. Chen, F. Zhou, X. Wu, Y. T. Chen, Y. Wang, F. Xue, J. C. Lee, *Physical Sciences Reviews* **1**, 5 (2016) (DOI: https://doi.org/10.1515/psr-2016-0011)
6. Y. C. Chen, Y. F. Chang, X. Wu, M. Guo, B. Fowler, F. Zhou, C. H. Pan, T. C. Chang, J. C. Lee, *ECS Transactions* **72**, 25 (2016) (DOI: 10.1149/07202.0025ecst).
7. K. C. Chang, T. C. Chang, T. M. Tsai, R. Zhang, Y. C. Hung, Y. E. Syu, Y. F. Chang, M. C. Chen, T. J. Chu, H. L. Chen, C. H. Pan, C. C. Shih, J. C. Zheng, and S. M Sze, *Nanoscale Research Letters* **10**, 1 (2015). (DOI: 10.1186/s11671- 015-0740-7)
8. T. J. Chu, T. C. Chang, T. M. Tsai, H. H. Wu, J. H. Chen, K. C. Chang, T. F. Young, K. H. Chen, Y. E. Syu, G. W. Chang, Y. F. Chang, M. C. Chen, J. H. Lou, J. H. Pan, J. Y. Chen, Y. H. Tai, C. Ye, H. Wang, and S. M. Sze, *Electron Device Letters, IEEE* **34**, 502 (2013). (DOI: 10.1109/LED.2013.2242843)
9. Y. C. Chen, Y. F. Chang, X. Wu, F. Zhou, M. Guo, C. Y. Lin, C. C. Hsieh, B. Fowler, T. C. Chang, and J. C. Lee, *RSC Advances* **7**, 12984 (2017). (DOI: 10.1039/C7RA00567A)
10. C. C. Hsieh, A. Roy, Y. F. Chang, D. Shahrjerdi, S. K. Banerjee, *Applied Physics Letters* **109** (22), 223501 (2017) (DOI: http://dx.doi.org/10.1063/1.4971188).
11. C. C. Hsieh, A. Roy, A. Rai, Y. F. Chang, S. K. Banerjee. *Appl. Phys. Lett.* **106**, 173108 (2015). (DOI: 10.1063/1.4919442)
12. C. C. Hsieh, A. Roy, Y. F. Chang, A. Rai, S. Banerjee, 73rd Annual Device Research Conference (DRC), 2015, 101 (2015) (DOI: 10.1109/DRC.2015.7175575).
13. C. C. Hsieh, Y. F. Chang, Y. Jeon, A. Roy, D. Shahrjerdi, S. K. Banerjee. IEEE *Electron Device Letters*, **38**, 871 (2017) (DOI: 10.1109/LED.2017.2710955).
14. Y. F. Chang, L. W. Feng, C. W. Huang, G. Y. Wu, C. H. Chang, J. J. Wu, S. Y. Wang, T. C. Chang, and C. Y. Chang, 2011 IEEE 4th International Nanoelectronics Conference, 1, (2011). (DOI: 10.1109/INEC.2011.5991628)
15. S. Kim, Y. F. Chang, B. G. Park, *RSC Advances* **7**, 17882-17888 (2017) (DOI: 10.1039/C6RA28477A).

16. S. Kim, Y. F. Chang, M. H. Kim, T. H. Kim, Y. Kim, B. G. Park, *Materials* **10**, 459 (2017) (DOI:10.3390/ma10050459).

17. S. Kim, Y. F. Chang, M. H. Kim, B. G. Park, *Appl. Phys. Lett.* **11**, 033509 (2017) (DOI: 10.1063/1.4985268).

18. L. W. Feng, C. Y. Chang, T. C. Chang, C. H. Tu, P. S. Wang, Y. F. Chang, M. C. Chen, H. C. Huang, *Appl. Phys. Lett.* **95**, 262110 (2009). (DOI: 10.1063/1.3279131)

19. C. C Cheng, C. H. Chien, G. L. Luo, J. C. Liu, Y. C. Chen, Y. F. Chang, S. Y. Wang, C. C. Kei, C. N. Hsiao, and C. Y. Chang, *J. Vac. Sci. Technol. B* **27**, 130 (2009). (DOI: 10.1116/1.3058724)

20. L. Ji, M. D. McDaniel, L. Tao, X. Li, A. B. Posadas, Y. F. Chang, A. Demkov, J. G. Ekerdt, D. Akinwande, R. S. Ruoff, J. C. Lee, and E. T. Yu, Tech. Dig. - Int. Electron Devices Meet. 8.6. 1-8.6. 3 (2014). (DOI: 10.1109/IEDM.2014.7047013)

21. F. Xue, A. Jiang, Y. T. Chen, Y. Wang, F. Zhou, Y. F. Chang, J. C. Lee IEEE Transactions on Electron Devices, **61**, 2332 (2014).
(DOI: 10.1109/TED.2014.2320946)

22. F. Zhou, F. Xue, Y. F. Chang, J. C. Lee, Device Research Conference (DRC), 2014 72nd Annual, 207 (2014). (DOI: 10.1109/DRC.2014.6872370)

23. F. Xue, A. Jiang, Y. T. Chen, Y. Wang, F. Zhou, Y. F. Chang, and J. Lee, Tech. Dig. - Int. Electron Devices Meet. 27.5. 1-27.5. 4 (2012).
(DOI: 10.1109/IEDM.2012.6479116)

24. Y. Wang, Y. T. Chen, F. Xue, F. Zhou, Y. F. Chang, and J. C. Lee, *ECS Transactions* **50**, 151 (2013). (DOI: 10.1149/05004.0151ecst)

25. M. Q. Guo, Y. C. Chen, C. Y. Lin, Y. F. Chang, B. Fowler, Q. Q. Li, J. Lee, Y. G. Zhao, *Applied Physics Letters* **110** (23), 233504 (2017)
(DOI: 10.1063/1.4985070).

26. C. Y. Lin, Y. C. Chen, M. Guo, C. H. Pan, F. Y. Jin, Y. T. Tseng, C. C. Hsieh, X. Wu, M. C. Chen, Y. F. Chang, F. Zhou, B. Fowler, K. C. Chang, T. M. Tsai, T. C. Chang, Y. Zhao, S. M Sze, S. Banerjee, J. C Lee, 2017 International VLSI Technology, Systems and Application (VLSI-TSA), 1, (2017) (DOI: 10.1109/VLSI-TSA.2017.7942474).

27. M. Guo, Y. C. Chen, Y. F. Chang, X. Wu, B. Fowler, Y. Zhao, J. C. Lee Device Research Conference (DRC), 2016 74th Annual, 1-2 (2016)
(DOI: 10.1109/DRC.2016.7548460).

28. C. Y. Lin, P. H. Chen, T. C. Chang, K. C. Chang, S. Zhang, T. M. Tsai, C. H. Pan, M. C. Chen, Y. T. Su, Y. T. Tseng, Y. F. Chang, Y. C. Chen, H. Huang, S. M. Sze, *Nanoscale*, **9** 8586 (2017) (DOI: 10.1039/C7NR02305G)

29. F. Zhou, L. Guckert, Y. F. Chang, E. E. Swartzlander Jr, J. C. Lee, *Appl. Phys. Lett.* **107**, 183501 (2015) (DOI: 10.1063/1.4934835).

30. Y. F. Chang, P. Y. Chen, B. Fowler, Y. T. Chen, F. Xue, Y. Wang, F. Zhou, J. C. Lee, *International Symposium on VLSI Technology, Systems, and Applications (VLSI-TSA)*. 1 (2013). (DOI: 10.1109/VLSI-TSA.2013.6545589)

31. Y. F. Chang, Y. C. Chen, J. Li, F. Xue, Y. Wang, F. Zhou, B. Fowler, J. C. Lee. *Device Research Conference (DRC), 71st Annual*. 135 (2013). (DOI: 10.1109/DRC.2013.6633830)

32. Y. Wang, Y. T. Chen, F. Xue, F. Zhou, Y. F. Chang, B. Fowler, J. C. Lee. *Appl. Phys. Lett.* **100**, 083502 (2012). (DOI: 10.1063/1.3687724)

33. Y. T. Chen, B. Fowler, Y. Wang, F. Xue, F. Zhou, Y. F. Chang, J. C Lee. *ECS Journal of Solid State Science and Technology* **1**, P148 (2012). (DOI: 10.1149/2.013203jss)
34. Y. C. Chen, H. C. Huang, C. Y. Lin, S. Kim, Y. F. Chang, J. C. Lee, *ECS Journal of Solid State Science and Technology*, **7**(8), P350-P354. (2018)
35. Y.C. Chen, Y.F. Chang, C.Y. Lin, X. Wu, G. Xu, B. Fowler, T.C. Chang, and J.C. Lee, *ECS Transactions*, *80*(10), pp.923-931 (2017)
36. Y.C. Chen, C.Y. Lin, H.C. Huang, S. Kim, B. Fowler, Y.F. Chang, X. Wu, G. Xu, T.C. Chang, and J.C. Lee, *Journal of Physics D: Applied Physics*, **51**(5), p.055108 (2018)
37. Y. C. Chen, X. Wu, Y. F. Chang, J. C. Lee, 76th IEEE. Device Research Conference (DRC), pp. 1-2, (2018)
38. Y.C. Chen, S.T. Hu, C.Y. Lin, B. Fowler, H.C. Huang, C.C. Lin, S. Kim, Y.F. Chang, and J.C. Lee, *Nanoscale*, **10**(33), pp.15608-15614 (2018).

Chapter 3

Poster Session

ECS Transactions, 89 (3) 55-59 (2019)
10.1149/08903.0055ecst ©The Electrochemical Society

Novel Exploration of Flat-Band Voltage Manipulation by Nitrogen Plasma Treatment for Advanced High-k/Metal-Gate CMOS Technology

Jiaxin Yao[a,b], Zhaozhao Hou[a,b], Zhenhua Wu[a,b], and Huaxiang Yin[a,b]

[a] Key Laboratory of Microelectronics Devices and Integrated Technology, Institute of Microelectronics of Chinese Academy of Sciences, Chaoyang, Beijing 100029, People's Republic of China
[b] University of Chinese Academy of Sciences, Beijing 100049, People's Republic of China

This work presents a novel exploration of flat-band voltage (V_{FB}) manipulation by nitrogen plasma treatment (NPT) in the high-k/metal-gate (HKMG) Metal-Oxide-Semiconductor Capacitor (MOSCAP) for the scaling-down in further CMOS fabrication technology. Significant V_{FB} shifts were achieved from −220 mV to +60 mV and from +130 mV to +420 mV for P- and N-MOSCAPs, respectively, with different RF powers settings. One unique modulation mechanism can successfully address the difference of P- and N-device variation by the NPT process, and thus be used to adjust the threshold voltage of the MOSCAPs in P-/N-logic devices scaled beyond 7-nm technology node.

Introduction

Threshold voltage (V_{th}) control is crucial for the shrinking in CMOS technology (1). Conventional metal gate stack technology is confronted with the limited gap filling of gate trenches in the FinFET or GAA (gate-all-around) FET, which can ruin the resistance reductions and device performance (2,3). Therefore, novel exploration of V_{th} modulation is relevant, especially for the advanced CMOS HKMG (high-k/metal-gate) technology. In this work, we focus on the modulation of the nitrogen plasma treatment of the first capping layer of a HKMG stack for both P- and N-type MOSCAPs. The C-V and I-V characteristics are measured for the V_{FB} (flat-band voltage) shift capability. The unique physical analysis for the different modulation mechanism for P- and N-MOSCAPs are put forward. The in-house developed NPT process demonstrates the potential of effective V_{FB} modulation for the advanced HKMG technology.

Fabrication of CMOSCAPs with the NPT process

The fabrication of the CMOSCAPs is illustrated in Fig. 1 (a). Primary N/P type (100) silicon wafers without well doping were used as substrates for preparing the P-/NMOSCAP structures. The high-k dielectric (HfO_2) was deposited by atomic layer deposition (ALD) after an interfacial layer (IL) formed with chemical oxidation. On the ALD-HfO_2 layer (HK), the first TiN capping layer and work function metal (WFM) were deposited in sequence for both the P- and N-MOSCAP. Here, the first TiN capping layer was deposited by ALD. Then, TaN was deposited by ALD as etching stop layer (ESL) only for the P-MOSCAP. Subsequently, the WFM layers for P- and N-MOSCAPs were TiN, deposited by Chemical Vapor Deposition (CVD) and TiAlC, deposited by ALD, respectively. In addition, the second TiN layer (top TiN barrier layer) was deposited on the WFM only for the N-MOSCAP before the deposition of a gap filling metal of tungsten (W). The finally fabricated gate stacks of P- and N-MOSCAPs are: IL (interfacial layer)/HK/TiN (capping layer)/TaN (ESL, etching stop layer)/TiN (p-WFM)/W (gap filling layer) and IL/HK/TiN (capping layer)/TiAlC (n-WFM layer)/TiN (barrier layer)/W, respectively, as shown in Fig. 1b.

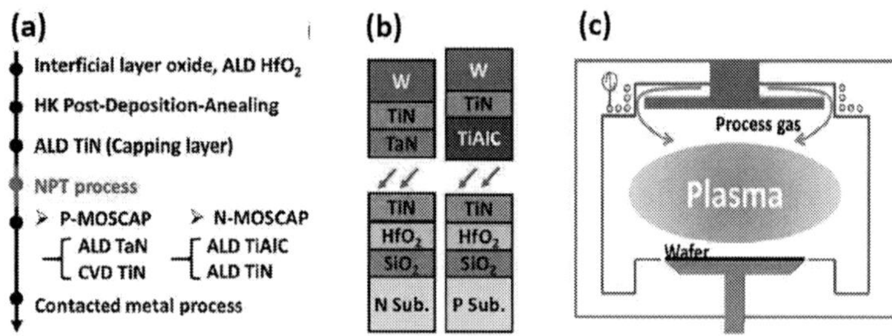

Figure 1. (a) Nitrogen plasma treatment (NPT) process flow with full gate-last CMOS compatible process. (b) Schematics of the P-MOSCAP HKMG stack (left) and the N-MOSCAP HKMG stack (right); the NPT process was applied on the first ALD-TiN capping layer. (c) Schematics of the plasma reactor.

To improve the film quality and recover the defects of the HfO$_2$ deposited by low-temperature ALD, the high-k dielectric film was submitted to a post-deposition anneal (PDA) at 450 °C for 15 secs in an N$_2$ atmosphere of 50 Torr.

To study the work function modulation of the gate stack in detail, the novel NPT process was performed under different conditions on the first ALD-TiN capping layer just after the first ALD-TiN layer deposition. The nitrogen plasma process chamber equipped with a remote inductively coupled plasma source was an Endura 5500 system produced by Applied Materials, Inc., see Fig. 1c. The wafer temperature was fixed at 400 °C and the chamber pressure was fixed at 1.3 Torr.

To systematically investigate the variation of the electrical properties by the NPT effect on the TiN layer, the radio frequency (RF) power of the nitrogen plasma source was set at 350, 550 and 750 W. The nitrogen flow ratio in the chamber was varied from 100 % to 40 % with a total flow-velocity of 750 sccm. The reduction of the nitrogen plasma ratio was accomplished by dilution with H$_2$ carrier gas. Since the generated nitrogen plasma density and ion bombardment energy is highly dependent on RF power and the nitrogen flow ratio, the NPT is expected to have a great effect on the electrical characteristics of MOSCAPs. More details on the experimental conditions for the NPT process can be found elsewhere (4).

To examine the gate modulation capability by NPT, the electrical characteristics of these fabricated MOSCAPs were investigated. The C-V and I-V curves were measured in a Keithley 4200 semiconductor characterization system in air ambient at room temperature. Several electrical parameters including V_{FB} and EOT were extracted by fitting the measured C-V data to a simulation program incorporated with the quantum mechanical effect correction developed by UC Berkeley (5). The gate leakage J_g was extracted from the I-V curve at (V_{FB}+1) V for the P-MOSCAP, and (V_{FB}-1) V for the N-MOSCAP, respectively.

Results and Discussion

The C-V and I-V characteristics are depicted in Fig. 2 (a)/(b), (c)/(d) for the P- and N-MOSCAPs, respectively. The C-V curves demonstrate the effective modulation by the NPT process both for the P- and the N-MOSCAPs. Compared to the samples without NPT, the gate leakage current of both P- and N-MOSCAPs are improved with the NPT process. The developed NPT process on the first ALD-TiN capping layer demonstrates significant V_{FB} modulation capability and causes a controllable effective work function of the metal gate: 1) shifting to band center with increasing radio frequency power both for P-/N-MOSCAPs; 2) shifting to opposite directions (P- to band edge, N- to band center) with reduced nitrogen flow ratio for P-/N-MOSCAPs; 3) shifting from -220 mV to +60 mV and from +130 mV to +420mV for the P- N-MOSCAPs and the N-MOSCAPs, respectively.

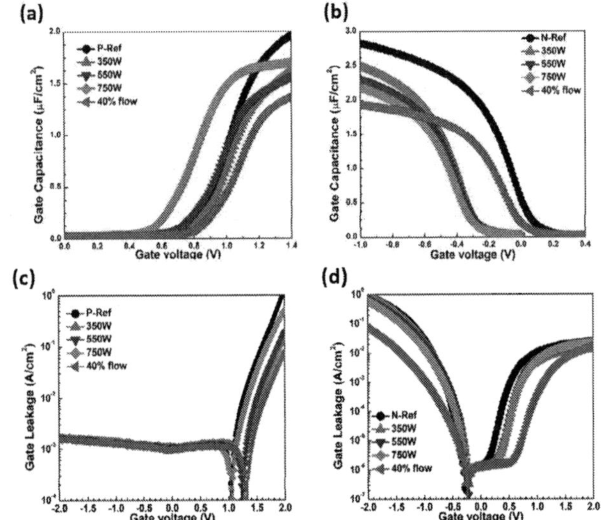

Figure 2. C-V characteristics for (a) P-MOSCAPs and (b) N-MOSCAPs;
I-V characteristics for P-MOSCAPs (c) and N-MOSCAPs (d).

Based on the different gate stacks of P-/N-MOSCAPs and the plasma effects, one unique physical model including N-vacancy/Al-trap effect is successfully proposed to illustrate the complicated shift behaviors both for HKMG P-/N-MOSCAPs, as demonstrated in Fig. 3 and described in detail in ref. 4. For the P-MOSCAP, the V_{FB} shift is mostly determined by N-vacancies, because the first capping layer (TiN) is the same as the work-function layer (TiN). Therefore, with the nitrogen vacancies increasing, V_{FB} demonstrates a negative shift towards the band center. For the N-MOSCAP, the V_{FB} shift is determined by both N-vacancies and Al-traps effect, because the first capping layer

(TiN) is different from the upper work-function layer (TiAlC). Therefore, nitrogen-vacancies can activate the Al-trap effect by the NPT applied in the first capping layer (TiN), where Al atoms from the upper WFM layer (TiAlC) movement are captured.

The model is further illustrated in Fig. 4.

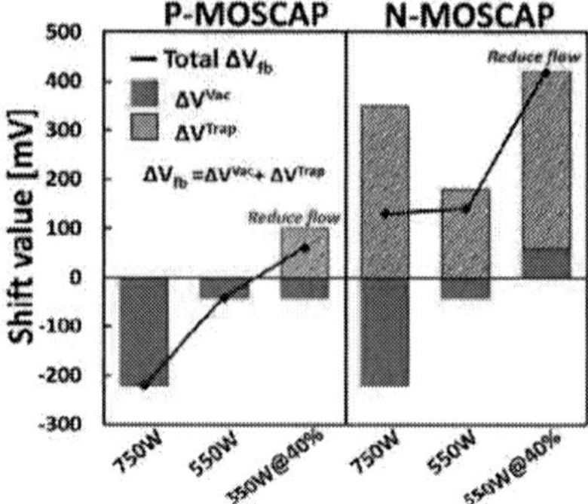

Figure 3. V_{FB} shift behavior via N-vacancy and Al-trap effects for both P- and N-MOSCAP. Zero represents the sample without NPT process. From ref. (4).

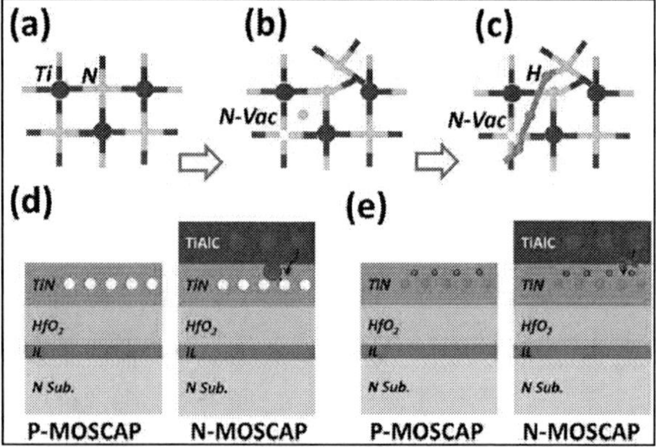

Figure 4. Illustrations for V_{FB} shifts by the NPT process under different RF-power and N_2 flow ratio conditions. (a) Surface TiN crystallinity before plasma treatment. (b) Decreased surface TiN crystallinity and N-vacancies appearing after NPT. (c) TiN crystallinity with reduced N_2 flow ratio. (d) N-vacancy effect and Al-trap effect in TiN surface with RF power condition. (e) N-vacancy and Al-trap block in the TiN surface with reduced N_2 flow ratio condition. From ref. (4).

Conclusions

In conclusion, the technology of NPT in the first ALD-TiN capping layer of HKMG MOSCAPs for gate work function modulations has been systematically investigated. Significant V_{FB} shift values, such as -220mV/+130mV with different RF powers and +60mV/+420mV with different nitrogen flow ratios have been achieved for P-/N-MOSCAPs, respectively. One simple theoretical model by employing the function of N-vacancies in the ALD-TiN capping layer with the corresponding Al-trapping effect by different WFM atoms in the metal-gate is presented, which gives a unique physical explanation to the complex V_{FB} adjustment behavior with different device types and process conditions. The results demonstrate that the NPT process is one of the most promising technologies for the V_{th} manipulation in future scaling CMOS technology nodes with less process cost.

Acknowledgments

The authors would like to thank all the people at the Integrated Circuit Advanced Process Center, Institute of Microelectronics, CAS, Beijing, China, for their assistance in the fabrication.

References

1. *2015 International Technology Roadmap for Semiconductors (ITRS)*, https://www.semiconductors.org/resources/2015-international-technology-roadmap-for-semiconductors-itrs/
2. J. Zhang, *et al.*, in *"2017 IEEE International Electron Devices Meeting"*, p. 22. 1.1 - 22.1.4 (2017).
3. N. Yoshida, *et al.*, in *"2017 IEEE International Electron Devices Meeting"*, p. 22.2.1 - 22.2.4 (2017).
4. J. Yao, H. Yin, Z. Wu, J. Gao, Q. Zhang, Z. Hou, J. Gu, and K. Luo, *ECS J. Solid State Sci. Technol.* 7(8), Q152-Q158 (2018).
5. http://www-device.eecs.berkeley.edu/qmcv/

Chapter 4

Processes for Advanced Integrated Circuits

ECS Transactions, 89 (3) 63-69 (2019)
10.1149/08903.0063ecst ©The Electrochemical Society

Hollow cathode plasma (HCP) enhanced atomic layer deposition of silicon nitride (SiN_x) thin films using pentachlorodisilane (PCDS)

Su Min Hwang[a], Aswin L. N. Kondusamy[a], Qin Zhiyang[a], Harrison Sejoon Kim[a],
Xin Meng[b*], Jiyoung Kim[a], Byung Keun Hwang[c], Xiaobing Zhou[c], Michael Telgenhoff[c],
and Jeanette Young[c]

[a] *Department of Materials Science and Engineering, The University of Texas at Dallas, 800 West Campbell Road, Richardson, Texas 75080, USA.*
[b] *Department of Electrical Engineering, The University of Texas at Dallas, 800 West Campbell Road, Richardson, Texas 75080, USA.*
[c] *The Dow Chemical Company, 2200 W. Salzburg Road, Midland, Michigan 48686, USA.*
Currently at Lam Research Corporation, 11155 SW Leveton Drive, Tualatin, Oregon 97062, USA

In this work, effects of NH_3/N_2 and N_2-H_2/Ar plasma gases for the growth of PEALD SiN_x films using pentachlorodisilane (PCDS, HSi_2Cl_5) were studied using a hollow cathode PEALD system. At identical process conditions, the combination of PCDS and N_2−H_2/Ar plasma showed a relatively lower (approximately < 10 %) growth rate as compared to NH_3/N_2 plasma under a range of process temperatures (240−300 °C) whereas the wet etch resistance to HF acid was improved (> 1.6 nm/min, 500:1 HF). Using XPS and FTIR analysis, it was identified that a N_2−H_2/Ar gas mixture results in a Si-rich SiN_x film with less N−H_x bonds when compared to NH_3/N_2 mixture, thereby resulting in a decreased wet etch rate.

Introduction

Silicon nitride (SiN_x) films have drawn great attention due to their wide range of applications such as passivation layer, gate dielectric, spacer, charge trap layer, and diffusion barrier (1). Conventionally used for SiN_x deposition, LPCVD and PECVD have limitations in terms of conformal deposition and thickness scalability (2). Since plasma enhanced atomic layer deposition (PEALD) is expected to overcome these shortcomings, several research groups have reported SiN_x film deposition by using plasma (3). A number of studies on PEALD SiN_x process using various chlorine-containing silicon precursors have been reported due to their ease in synthesis as well as good thermal stability (4–8). Recently, PEALD SiN_x film growth using a novel chlorosilane precursor, pentachloro-disilane (PCDS, HSi_2Cl_5) with NH_3/N_2 plasma was reported (9). Using NH_3/N_2 plasma, PCDS enhanced the growth rate (approximately > 20 %) compared to hexachlorodisilane (HCDS, Si_2Cl_6), but still showed relatively poor wet etch resistance. Hence, we explore the feasibility of improving the wet etch resistance of PEALD-SiN_x films using a different gas mixture in plasma. It has been reported earlier, that the wet etch rate of SiN_x films has a linear relationship with the hydrogen concentration in the films (1,10–12). Compared to NH_3/N_2 plasma, N_2−H_2 (forming gas, FG) plasma improves the wet etch rate possibly due to less N−H and H−N−H bonds in the SiN_x films (8). Moreover, PEALD of SiN_x using a chlorosilane precursor and NH_3 free-plasma gas has been rarely reported.

In this work, we have investigated the wet etch rate of PEALD SiN_x films deposited using PCDS as the Si precursor and different plasma gases (either NH_3/N_2 or FG/Ar) with a hollow cathode PEALD system. In comparison to the SiN_x films under the NH_3/N_2 plasma condition, the combination with PCDS and N_2–H_2 plasma shows a higher wet etch resistance to HF acid (1.6 nm/min, 500:1 HF). In addition, X-ray reflectivity (XRR), X-ray photoelectron spectroscopy (XPS), and Fourier transform infrared spectroscopy (FTIR) were employed to characterize the SiN_x films.

Experimental details

PEALD SiN_x films were deposited using a home-built PEALD system as shown in Figure 1 (a). A hollow cathode plasma source (Meaglow Ltd.) connected to a 13.56 MHz RF generator was installed approximately 5 inches above the substrate holder. For the PEALD SiN_x process, PCDS (provided by The Dow Chemical Company) and a mixture of either NH_3/N_2 or forming gas (FG, 10% H_2–90% N_2)/Ar were used as the silicon and nitrogen source, respectively. The precursor delivery lines and exhaust tube were heated to 120 °C to avoid the condensation of precursors or residual products. 4-inch p-type silicon (100) wafers (Silicon Valley Microelectronics) were used as substrates. Each wafer was cleaned with 100:1 HF (49 % $HF:H_2O$, 1:100) to remove the native oxide, followed by loading into the PEALD chamber. The chamber was pumped down to ~10^{-6} Torr using a turbomolecular pump. The process pressure was maintained at 0.45 Torr with a continuous flow of either N_2 or Ar carrier gas. A representative time sequence of one cycle of PEALD SiN_x process is illustrated in Figure 1 (b). During the first precursor half-cycle, PCDS was exposed for 1 s, followed by a precursor purging time of 30 s. During the nitrogen reactant half-cycle, either NH_3 or forming gas was introduced. After 5s for the pressure stabilization, the RF power was set to 100 W including the minimum reflected power (< 2 W) for 15 s, followed by a purging time of 30 s.

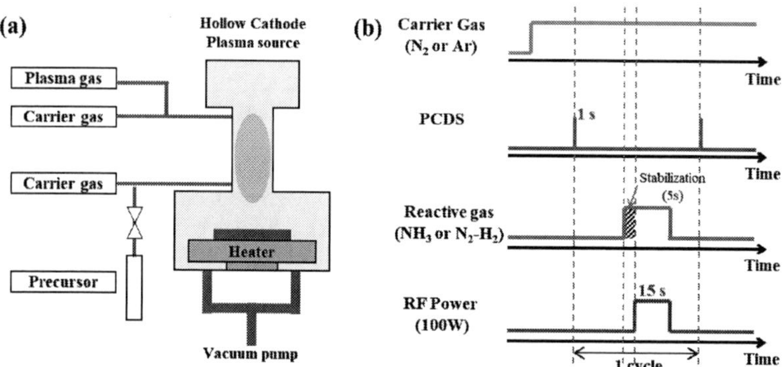

Figure 1. Schematic of (a) PEALD system and (b) representative time sequences of PEALD SiN_x process.

Spectroscopic ellipsometry (M-2000DI, J.A. Woolam) was used to measure the film thickness. The spectra at three different incident angles (55, 65, and 75 °) were collected and fitted using a Tauc-Lorentz model over 200–1600 nm wavelength range. To obtain the wet etch rates, each sample was dipped into a 500:1 dilute HF solution for 1–2 min, followed by rinsing with DI water and blown dried with N_2 gas. The difference of film thickness before and after etching was divided by the etching time to obtain the wet etch rate. To compare the film densities, X-ray reflectivity (XRR) using a Rigaku Smart Lab XRD system was performed. In the X-ray diffractometer, a 5.0 ° solar slit and a 5.0 mm slit were installed to incident and receiving parts, respectively. Rigaku GlobalFit software was employed to extract the film densities from the XRR scans. X-ray photoelectron spectroscopy (XPS, PHI VersaProbe II) with a monochromatic Al Kα X-ray source (1486.6 eV) was employed for film composition analysis. All film surfaces were pre-sputtered with Ar^+ ions (1 kV, 1 ×1 mm^2) for 2 min. The elemental composition of the films was estimated using peak areas and atomic sensitivity factors (13). A Fourier transform infrared (FTIR) spectroscopy (Thermo Electron Nicolet 4700) equipped with a Globar IR source and an MCT-A detector was employed to identify the bonding information. The infrared beam passes through a KBr beam splitter. The IR and background spectra were obtained in transmission mode using the deposited SiN_x film and HF-cleaned (100:1 diluted HF) sample from the same silicon wafer, respectively.

Results and discussion

<u>Investigation of the temperature dependence</u>

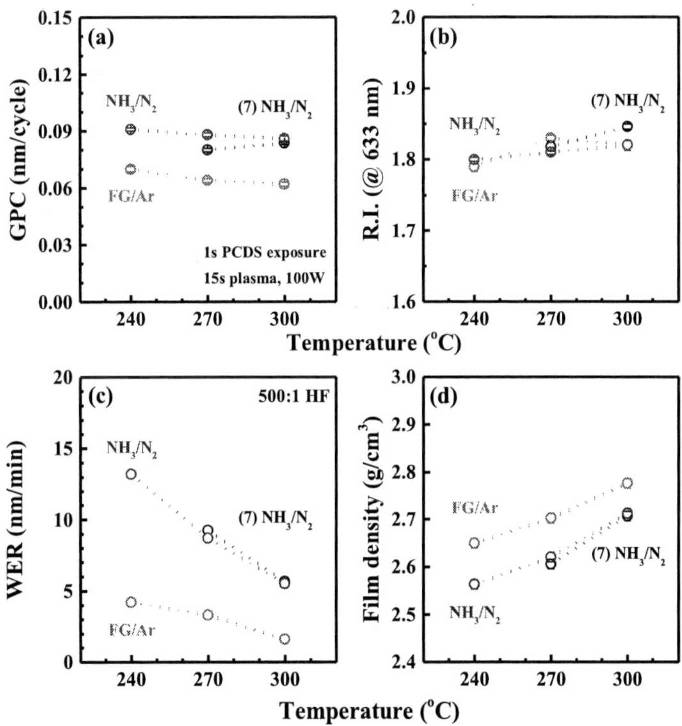

Figure 2. (a) Growth rate per cycle, (b) Refractive index, (c) WER, and (d) film density of deposited SiN_x films using PCDS precursors and different reactive gas mixtures at various temperatures.

The effect of plasma gas on the growth of SiN_x films was investigated by comparing the film properties such as growth rate (GPC), refractive index (R.I.), wet etch rate (WER), and film density in the temperature range 240–300 °C. The plasma gas compositions used in this study are NH_3/N_2 (25/50 sccm) and FG/Ar (10% H_2 balance N_2, 50/50 sccm). In Figure 2(a), both NH_3/N_2 and FG/Ar plasma gases show nearly constant growth rates within the temperature range of 240–300 °C. Compared to the previously reported growth rates (GPC of ~0.08 nm/cycle) using PCDS and under NH_3-rich condition, (NH_3/N_2, 90/30 sccm), the results under NH_3-deficient condition shows similar temperature dependence behavior. In addition, in a previous study we reported that PCDS based PEALD of SiN_x film exhibited a self-limiting growth behavior with 1s exposure (9). The results indicate that not only PCDS exhibits a self-limiting growth behavior, but also it is irrelevant to gas composition in case of the NH_3/N_2 plasma gas. Particularly, FG/Ar plasma gas, one of the candidates for the NH_3-free gas, can generate reactive NH_x species, which can react with chlorosilane (6,14–17). Therefore, it can be used in the PEALD SiN_x process as the NH_3-free plasma gas. On the other hand, FG/Ar results in approximately 25% lower growth rates (0.066± 0.004nm/cycle) than NH_3/Ar (0.088± 0.003nm/cycle). This difference in the growth rate could be attributed to the different amounts of reactive NH_x species from each plasma gas. During the plasma exposure, FG/Ar plasma gas provides fewer reactive species, affecting the reaction with the chemisorbed PCDS (9).

As the temperature increases from 240 to 300 °C, the R.I. values of both films increase from 1.79 to 1.83 (Figure 2(b)), whereas the WER significantly decreases (Figure 2(c)). This gradual decrease of wet etch rate at an elevated temperature can be explained by the reduction of the hydrogen concentration in the films. Harrison *et al.* recently reported that the hydrogen concentration of the SiN_x films decreased by increasing the process temperature, resulting in a lower WER in a diluted HF solution (1). On the other hand, FG/Ar gas shows improved wet etch resistance (1.6−4.2 nm/min), which are comparable to previously reported WER results and of SiN_x films deposited with chlorosilane precursors (7,18–20). As shown in Figure 2(d), the film density of both SiN_x films increases with increasing temperature (from 2.65 to 2.78 g/cm^3 for FG/Ar and from 2.56 to 2.71 g/cm^3 for NH_3/N_2). This could be related to the densification by the reduction of hydrogen concentration at an elevated temperature (21,22).

<u>Film characterization</u>

The composition of the SiN_x films deposited using different plasma gas mixtures at 270 °C was investigated by XPS analysis. As shown in Figure 3(a), both Si 2p and N 1s elemental scans clearly show peaks with a binding energy of ~101.8 eV and ~397.8 eV, respectively. These binding energies correspond to the SiN_x peak position (13). The absence of shoulder peak at ~99.6 eV, assigned to Si-Si dimer, indicates that Si−Si ligands from PCDS fully dissociate into Si dangling bonds after plasma gas exposure. In addition, both SiN_x films contain oxygen impurity less than 4 at. %. The low base pressure (~10^{-6} Torr) before the process and applying the hollow cathode plasma source can effectively decrease the oxygen contaminants such as H_2O and O_2 (23). In Cl 2p narrow scan, distinguishable tiny peaks were observed in both plasma gas conditions. These peaks correspond to incomplete Si−Cl bonds from PCDS, which is due to their higher bond dissociation energy (24,25). Furthermore, FG/Ar plasma gas, containing fewer reactive NH_x species than NH_3/N_2 plasma gas, results in insufficient ligand exchanges with a Si-Cl terminated surface. Consequently, the peak area of Cl 2p spectra accounts for approximately 3.0 and 1.5 at. %, respectively.   By comparing the elemental

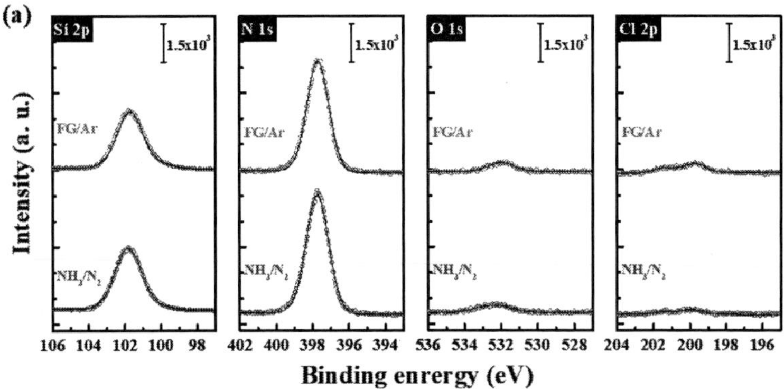

Figure 3. (a) XPS spectra of PEALD SiN$_x$ film deposited with different reactive gas mixtures at 270 °C and (b) elemental composition calculated from XPS spectra.

	[Si] at. %	[N] at. %	[O] at. %	[Cl] at. %	[C] at. %	[N] / [Si]
NH₃/N₂	44.9	48.9	3.2	3.0	< d.l.	1.09
FG/Ar	44.6	50.2	3.8	1.5	< d.l.	1.13

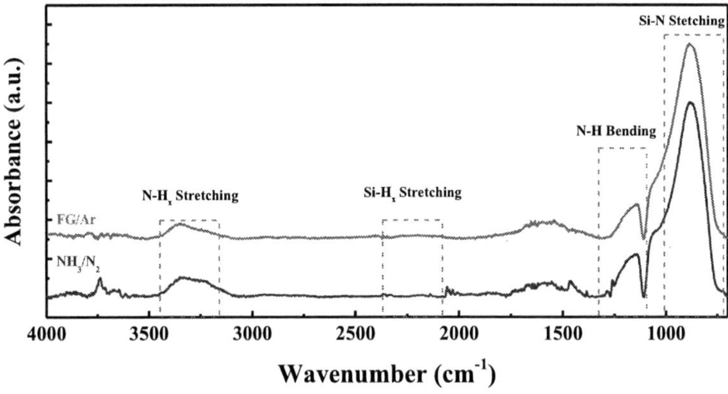

Figure 4. FTIR spectra of PEALD SiN$_x$ films deposited with NH₃/N₂ (blue) and FG/Ar(red) at 270 °C.

composition of the two films (Figure 3(b)), the SiN$_x$ film deposited using FG/Ar plasma is relatively Si-rich with N/Si ratios of 1.09 compared to case of NH₃/N₂ (1.13). During the FG/Ar plasma gas discharging, the generation of reactive nitrogen and hydrogen species (e.g., N_2^+, N^+, H_2^+, and H^+, and etc.) is dominant over NH$_x$ species (16). Therefore, NH$_x$ deficient plasma reacts less intensively with the chemisorbed PCDS, leading to relatively Si-rich films.

The normalized FTIR spectra are plotted in the range of 700–4000 cm^{-1} to identify the bonding information of each SiN$_x$ film. In Fig. 4, both plasma gases clearly show a peak at ~870 cm^{-1} corresponding to the Si−N stretching mode, whereas the Si−H stretching mode peak at ~2150 cm^{-1} is indistinguishable (11,20,26). This implies that the original

Si–H bonds from PCDS are involved in the surface reaction, and eventually the formation of Si–N bonds via ligand exchange occurs during the first half-cycle (precursor exposure step) or the following plasma exposure step. The bands at ~1180 cm^{-1} and ~3340 cm^{-1} are attributed to the N–H bending mode and N–H_x ($x = 1, 2$) stretching mode in the SiN_x films, respectively (27). For FG/Ar plasma, both the N–H bending mode and N–H_x stretching mode decreases compared to NH_3/N_2. Particularly, in the N–H_x stretching mode, the non-symmetrical NH_2 stretching mode at ~3250 cm^{-1} decreases significantly, resulting in an overall reduction of absorbance by approximately ~38 %. This curtailment in hydrogen concentration can be attributed to the fewer reactive NH_x species during the plasma discharging.

Based on Figures 3 and 4, the proposed reaction pathway using PCDS and FG/Ar plasma gas is similar to previous studies with chlorosilane precursor and NH_3 plasma gas (14,26,28). Meng *et al.* reported that PCDS is chemisorbed on the surface terminated with amine groups (−NH_2 and −NH−) by liberating either H_2 or HCl as the byproducts, eventually forming the SiN_x embedded with hydrogen bonds (9). In the present study, the surface after FG/Ar plasma exposure is terminated with either bare =N− or −NH− groups due to the lack of reactive NH_2 species compared to NH_3/N_2 plasma. This different surface termination affects both the following PCDS chemisorption and the reaction with HF acid, consequently, it shows relatively lower WER values compared to NH_3/N_2. On the other hand, the reactive hydrogen species in the FG/Ar partially change the surface termination from uncoordinated =N− to −NH_x, resulting in the further reaction with PCDS. This is in agreement with the higher growth rates than those reported in previous studies using chlorosilane and N_2 plasma (9,28). Therefore, based on the proposed reaction pathway and our experimental analysis, we suggest that FG/Ar plasma gas can be applied to PEALD SiN_x with chlorosilane precursors, to reduce the hydrogen concentration in the film.

Conclusions

We have demonstrated the hollow cathode PEALD of SiN_x films using PCDS and NH_3/N_2 or FG/Ar in the temperature range of 240−300 °C. We found out that the combination of the chlorine-based precursor and FG/Ar (NH_3-free gas) gas plasma is applicable to PEALD SiN_x process. In comparison to the SiN_x films deposited with NH_3/N_2 plasma, the SiN_x films deposited using the FG/Ar plasma gas showed a higher wet etch resistance and higher film density under identical process conditions whereas the GPC decreased by approximately 10%. We have found that the FG/Ar plasma contributes to a lower N–H_x bond concentration on the surface as well as to a partial surface treatment, resulting in curtailment of the wet etch rate in HF acid.

Acknowledgments

We thank The Dow Chemical for providing the financial support for this work and for providing the PCDS, as well as for technical discussions. The authors would like to acknowledge Dr. Scott Butcher (Meaglow Ltd.) for the technical discussions on the use of the hollow cathode plasma source.

References

1. H. S. Kim et al., *ACS Appl. Mater. Interfaces*, **10**, 44825–44833 (2018).
2. J. Kim and T. W. Kim, *JOM*, **61**, 17–22 (2009).
3. X. Meng et al., *IEEE Electron Device Lett.*, **39**, 1195–1198 (2018).
4. A. Nakajima, Q. D. M. Khosru, T. Yoshimoto, T. Kidera, and S. Yokoyama, *Appl. Phys. Lett.*, **80**, 1252–1254 (2002).
5. S. Morishita, S. Sugahara, and M. Matsumura, *Appl. Surf. Sci.*, **112**, 198–204 (1997).
6. R. A. Ovanesyan, D. M. Hausmann, and S. Agarwal, *ACS Appl. Mater. Interfaces*, **7**, 10806–10813 (2015).
7. K. Park et al., *Thin Solid Films*, **517**, 3975–3978 (2009).
8. X. Meng et al., *Materials*, **9**, 1007 (2016).
9. X. Meng et al., *ACS Appl. Mater. Interfaces*, **10**, 14116–14123 (2018).
10. R. Chow, W. A. Lanford, W. Ke-Ming, and R. S. Rosler, *J. Appl. Phys.*, **53**, 5630–5633 (1982).
11. G. N. Parsons, J. H. Souk, and J. Batey, *J. Appl. Phys.*, **70**, 1553–1560 (1991).
12. W. A. Lanford and M. J. Rand, *J. Appl. Phys.*, **49**, 2473–2477 (1978).
13. J. Moulder, W. Stickle, P. Sobol, and K. Bomben, *Handbook of X-ray Photoelectron Spectroscopy* J. Chastain, Editor, 2nd ed., Enen Prairie, (1992).
14. C. A. Murray, S. D. Elliott, D. Hausmann, J. Henri, and A. LaVoie, *ACS Appl. Mater. Interfaces*, **6**, 10534–10541 (2014).
15. R. Avni and T. Spalvins, *Mater. Sci. Eng.*, **95**, 237–246 (1987).
16. M. F. J. Vos, G. van Straaten, W. M. M. Kessels, and A. J. M. Mackus, *J. Phys. Chem. C*, **122**, 22519–22529 (2018).
17. Ö. Danielsson and E. Janzén, *J. Cryst. Growth*, **253**, 26–37 (2003).
18. S. Riedel, J. Sundqvist, and T. Gumprecht, *Thin Solid Films*, **577**, 114–118 (2015).
19. W.-J. Lee et al., *Korean J. Mater. Res.*, **14**, 90–93 (2010).
20. S. Yokoyama, N. Ikeda, K. Kajikawa, and Y. Nakashima, *Appl. Surf. Sci.*, **130–132**, 352–356 (1998).
21. T. Faraz et al., *ACS Appl. Mater. Interfaces*, **9**, 1858–1869 (2017).
22. P. C. Jhang et al., *Solid. State. Electron.*, **133**, 10–16 (2017).
23. C. Ozgit-Akgun, E. Goldenberg, A. K. Okyay, and N. Biyikli, *J. Mater. Chem. C*, **2**, 2123–2136 (2014).
24. C. Petit-Etienne et al., *J. Vac. Sci. Technol. B, Nanotechnol. Microelectron. Mater. Process. Meas. Phenom.*, **31**, 011201 (2012).
25. Y.-R. Luo, *Comprehensive Handbook of Chemical Bond Energies*, p. 1-1687, CRC Press, Boca Raton, (2007).
26. J. W. Klaus, A. W. Ott, A. C. Dillon, and S. M. George, *Surf. Sci.*, **418** (1998).
27. D. V. Tsu, G. Lucovsky, and M. J. Mantini, *Phys. Rev. B*, **33**, 7069–7076 (1986).
28. L. L. Yusup et al., *RSC Adv.*, **6**, 68515–68524 (2016).

ECS Transactions, 89 (3) 71-76 (2019)
10.1149/08903.0071ecst ©The Electrochemical Society

Co-Planar Structured Nano-Resistor Devices

Yue Kuo

Thin Film Nano & Microelectronics Research Laboratory, Texas A&M University, College Station, TX 77843-3122, USA

Co-planar structured nano-resistor devices made from co-planar MOS capacitors with a thin high-k gate dielectric have been fabricated and studied. The complete device was made of a 2-mask process with IC compatible materials. Nano-resistors were self-aligned to the gate electrode after a dielectric breakdown process. The broad band white light was emitted from the device upon the application of a gate voltage larger than the breakdown voltage due to the thermal excitation of nano-resistors. Optical characteristics of the co-planar structured device were compared with those of a non co-planar structured device. The main advantage of the co-planar structured device is that is can be driven without being interfered by adjacent devices. Therefore, a large number of device can be fabricated into an array for various electronic and optoelectronic applications.

Introduction

In the past half century, the advancement of the IC technology followed the spirit of the Moore's Law in reducing the device dimension into the nano-size in order to maximizing the packing density into a chip (1). The fabrication of nano-devices usually requires two critical process steps: the nano-sized pattern definition and the low undercut etching. The former requires a deep UV, e-beam, or non-conventional lithography process; the latter is commonly carried out using a low etch-bias reactive ion etching (RIE) process. Both processes depend on the availability of very expensive and delicate equipment. Some novel methods have been proposed to prepare nano devices without requiring the high-priced equipment, such as the selective growth or formation of nano-rods or nano-dots (2). They are used in production because of practical concerns, such as throughput, yield, and reliability.

Resistors are widely used in various circuits or products. When other devices in a product are shrunk to the nano-size, the dimension of the resistor needs to be reduced accordingly. Most nano-resistors are of the nano-wire shape horizontally or vertically aligned to the substrate surface (3,4). In general, nano-wires are easily influenced by the environment, such as quick oxidation in air even at room temperature. Kuo's group reported a novel type of nano-resistor devices, as shown in Figure 1 (5,6,7). Each nano-resistor is surrounded by a high quality high-k dielectric. The nano-resistor device has a linear current density (J) vs. voltage (V) relationship when the magnitude of the applied voltage is larger than that of a threshold voltage, as shown in Figure 2.

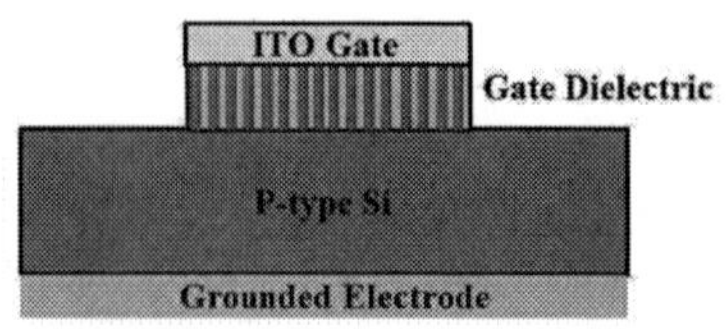

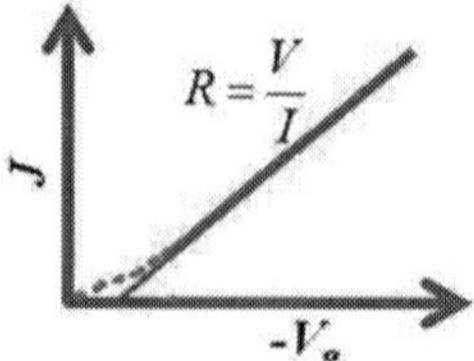

Fig. 1. A nano-resistor device.   Fig. 2. *J-V* relationship of a nano-resistor device.

These nano-resistors are self-aligned to the top gate electrode that can be made of a metal, a transparent conductor, such as ITO, or a heavily doped Si thin film. They are formed from the breakdown of a MOS capacitor with a high-*k* gate dielectric layer on a Si wafer (5). The nano-resistor device has many interesting properties suitable for electronic and optoelectronic applications. For example, it is named a solid-state incandescent light emitting device (SSI-LED) because light is emitted from the thermal excitation of the nano-resistors upon the passage of a large current (5,6,7). The emitted light covers the whole visible and part of the near IR wavelength range similar to that of the conventional incandescent light bulb. The light spectrum is located in the warm white light region of the CIE chart with a large CCT (color correlated temperature) of above 2,500 K and a high CRI (color rendering index) of > 95. Since the material and fabrication process of the nano-resistor device are completely IC compatible, the cost of the device is very low. In addition, the wavelength of the broadband light can be narrowed down to near 850 nm using a thin film filter (8,9). When used in combination with the proper waveguide and modulation units, the SSI-LED can be a light source for the on-chip optical interconnect. Separately, the nano-resistor device can be used as a diode or antifuse in low- or medium-power ICs (10).

Upon the application of a voltage on the nano-resistor device, shown in Fig. 1, upon the application of a voltage, the current flows perpendicular to the wafer surface, i.e., between the gate electrode on the wafer surface and the electrode on the back of the wafer. All nano-resistor devices on the same wafer have to share the same bottom electrode. This causes serious interference among adjacent devices, and limits many practical applications. In this paper, we present a new co-planar structured nano-resistor device, which can be operated free of the interferences of the adjacent device. The electrical and optical characteristics of the device are shown and discussed.

Experimental

<u>Fabrication of Co-planar Structured Nano-resistors</u>

The co-planar MOS capacitor with the grounded electrode located on the same side of the gate electrode is fabricated, as shown in Figure 3(a). It is made with two mask steps. The ZrHfO gate dielectric and the ITO gate metal layers were deposited on a pre-cleaned p-type Si wafer. The first mask was used to define and etch the gate electrode. After the photoresist is reflowed, the gate dielectric layer was wet-etched. Then, a metal layer was sputter deposited. The metal layer on top of the photoresist was lifted off after the photoresist was stripped off. The second mask was used to define the grounded electrode. This process warrants the formation of a uniform gap between the gate and the grounded electrodes. With the application of a large negative gate voltage ($-V_g$) to the ITO gate, the gate dielectric is broken and the nano-resistors are formed, as shown in Fig. 3(b). A control sample of the vertical nano-resistor structure the same as that of Fig. 1 was also fabricated and characterized for comparison.

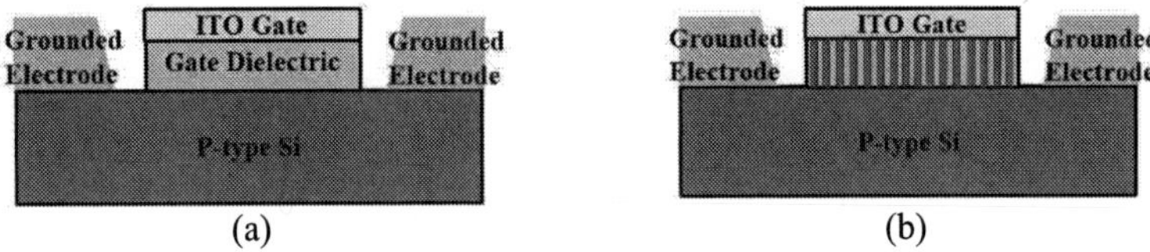

(a)   (b)

Fig. 3. (a) Designs of co-planar MOS capacitor and (b) co-planar nano-resistor.

<u>Electrical and Optical Characterization</u>

J-V curves of the capacitor before and after the dielectric breakdown were measured to determine the breakdown voltage (V_{BD}) at which conductive paths were formed. For the light emission study, a -V_g larger than that of the V_{BD} was applied and the emission spectrum was measured with an optical emission spectrometer (StellarNet BLK-C-SR-TEC) through an optical fiber. The spectrum of the co-planar SSI-LED was compared with that of the control sample.

Results and Discussion

<u>Formation of co-planar nano-resistor structure</u>

A top view image of the co-planar MOS capacitor is shown in Figure 4. A uniform gap space between the ITO and the grounded electrodes has been consistently formed from the lift off process. Properties of the capacitor were calculated from the *C-V* measurement. It had a low equivalent oxide thickness (EOT) of ~ 5 nm. Its electrical properties were as good as those of the control sample, e.g., a low interface density (D_{it}) of < 10^{11} cm^{-2}eV^{-1}, a small hysteresis flat band voltage shift (ΔV_{FB}) of ~ 0.06 V, and a low oxide trapping density (Q_{ot}) of ~ 10^{11} cm^{-2}.

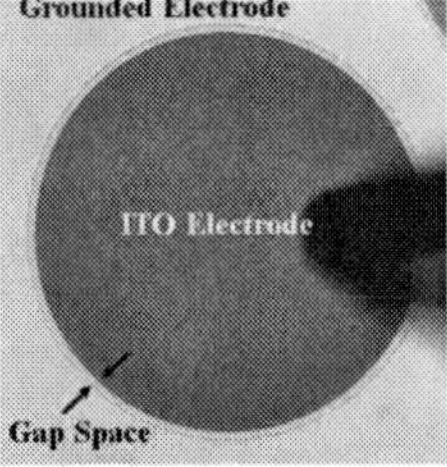

Fig. 4. Top view of the co-planar MOS capacitor.

The formation of the nano-resistors from the co-planar MOS capacitor could be verified from the abrupt jump of the current in the *J-V* curve. Figure 5 shows *J-V* curves before and after dielectric breakdown of the (a) control and (b) co-planar MOS capacitors. The large leakage current after the breakdown indicates the irreversible breakage of the gate dielectric. Since the current flow path in the co-planar structure is much smaller than that in the control sample, i.e., 8 μm vs. 500 μm Si paths, the voltage drop through the high-*k* dielectric layer probably dominated the dielectric breakdown process.

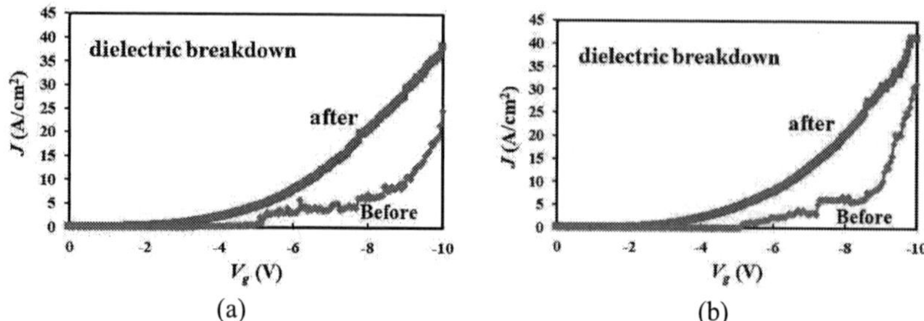

(a)   (b)

Fig. 5. J-V curves of (a) control and (b) co-planar MOS capacitors before and after dielectric breakdown.

<u>Light emission from nano-resistors</u>

The formation of nano-resistors from the co-planar capacitor was confirmed from the light emission pattern. Figure 6 shows the magnified micrographs of the (a) control and (b) co-planar devices stressed at $V_g = -30$ V. In both cases, light is emitted from individual dots uniformly distributed across the ITO gate electrode except for a higher density near the edge. Each light dot corresponds to a nano-resistor that was thermally excited due to the passage of the current. The number of light dots in the co-planar sample appears to be higher than that in the control sample.

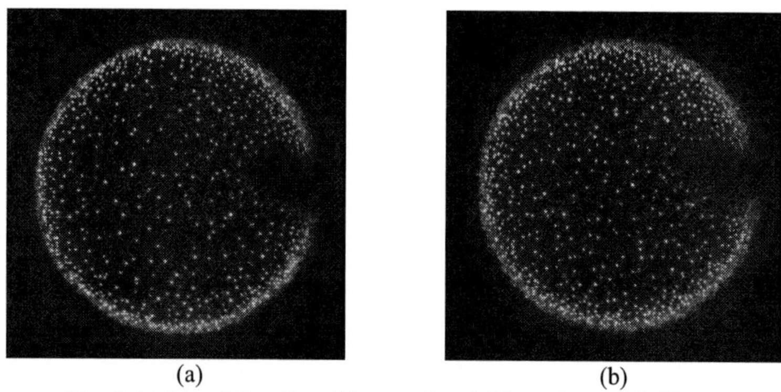

(a)   (b)

Fig. 6. Light emitting from (a) control and (b) co-planar SSI-LEDs.

<u>Emission spectra of nano-resistor devices</u>

Figure 7 shows the normalized emission spectra of the control and co-planar devices. The overlap of these spectra indicates that the composition and structure of nano-resistors in these 2 samples are probably the same. Their other optical properties are similar, such as for the CCT ~ 2,500K and the CRI ~ 98.

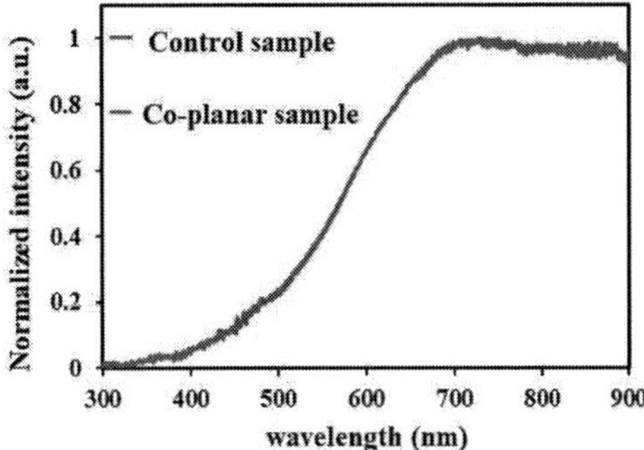

Fig. 7. Normalized emission spectra of control and co-planar SSI-LEDs.

Currently, it is difficult to compare the light emission efficiencies of the control and the co-planar SSI-LEDs because of the lack of an accurate measurement method. The co-planar device has about the same efficiency as that of the control sample when the external quantum efficiency (*EQE*) of the device was estimated by dividing the number of emitted photons (*NP*) with the total number of electrons provided to the device, as shown in the following equations (11)

$$EQE = \frac{NP}{I/q} \tag{1}$$

$$NP = \frac{2\pi r^2}{A_{slit}} \int \frac{E\,(\lambda)d\lambda}{hc/\lambda} \tag{2}$$

where I is the current in amps, q is the charge of an electron, $2\pi r^2/A_{slit}$ is the factor of fraction of light collected by an entrance slit, $E(\lambda)$ is the irradiance of the light source, λ is the light wavelength, h is the Planck constant, and c is the speed of light.

Conclusion

In summary, a new type of co-planar structured nano-resistor device has been fabricated using a 2-mask process and IC compatible materials. Nano-resistors, which were self-aligned to the gate electrode, were formed from the dielectric breakdown of a co-planar MOS capacitor containing a nm-thick high-*k* gate dielectric film. Light was emitted from individual nano-resistors evenly across the ITO gate electrode except for a slightly higher density near the edge. Optical characteristics and the *EQE* of the co-planar device were the same as those of the control sample reported in previous papers. Since each co-planar device can be driven individually free of the interference of the adjacent device, a large number of this kind of devices can be fabricated into an array for various electronic and optoelectronic applications.

Acknowledgments

Shumao Zhang is acknowledged for samples fabrication and characterization.

References

1. *IEEE Spectrum Special Report*: 50 Years of Moore's Law, http://spectrum.ieee.org/static/special-report-50-years-of-moores-law
2. W. I. Park, D. H. Kim, S.-W. Jung, and G.-C. Yi, *Appl. Phys. Lett.*, **80**, 4232 (2002).
3. N. Misra, L. Xu , Y. Pan, N. Cheung, and C. P. Grigoropoulos, *Appl. Phys. Lett.*, **90**, 111111 (2007).
4. S. Lee, B. Park, J. S. Kim, and T.-i. Kim, *Nanotech.*, **27**, 474001 (2016).
5. Y. Kuo and C.-C. Lin, *Appl. Phys. Lett.*, **102**, 031117 (2013).
6. Y. Kuo and C.-C. Lin, *Solid State Electrons*, **89**, 120 (2013).
7. Y. Kuo, *IEEE Trans. Electron Devices*, **62**, 3536 (2015).
8. S. Radovanovic, A.-J. Annema, and B. Nauta, *High-Speed Photodiodes Standard CMOS Technol.*, p. 24 (2006).
9. S. Zhang and Y. Kuo, *ECS J. Solid State Sci. Technol.*, **6**, Q39 (2017).
10. Y. Kuo, *ECS Trans.*, **69**(12), 23 (2015).
11. B. S. Harrison, Ph. D. Dissertation, University of Florida, p. 139 (2003).

Chapter 5

MOL/BEOL Material and Process Technology

ECS Transactions, **89** (3) 79-86 (2019)
10.1149/08903.0079ecst ©The Electrochemical Society

Etch Damage Reduction of Ultra Low-k Dielectric by Using Pulsed Plasmas

J. K. Jang [a,b], H. W. Tak [a], K. C. Yang [a], Y. J. Shin [a], J. Y. Hyeon [b], M. G. Kang [b]
and J. H. Ahn[b] and G. Y. Yeom[a,c]

[a] School of Advanced Materials Science and Engineering, Sungkyunkwan University,
2066 Seobu-ro, Jangan-gu, Suwon-si, Gyeonggi-do, Republic of Korea
[b] Foundry Division, Samsung Electronics, 1, Samsung-ro, Giheung-gu, Yongin-si,
Gyeonggi-do, Republic of Korea
[c] SKKU Advanced Institute of Nano Technology (SAINT), Sungkyunkwan University,
2066 Seobu-ro, Jangan-gu, Suwon-si, Gyeonggi-do, Republic of Korea

To reduce interconnect resistance and capacitance (RC) time delay
of semiconductor integrated circuit, the dielectric material with
more porosity is used in recent interconnection for lower dielectric
constant. However, it is difficult to use highly porous low-k
dielectric materials at the narrow pitch because it is easily damaged
during the plasma etching processes. In this study, as one of the
plasma induced damage reduction methods in the etching of
porous low-k dielectric using C_4F_8-based gases, RF pulsed plasma
methods have been investigated by using a dual frequency
capacitively coupled plasma etch system. RF pulsed plasmas
generated higher carbon-rich radicals in the plasma and less
charging effect on the dielectric material surface compared to
continuous wave plasmas and, therefore, showed a reduced
damaged layer compared to the conventional continuous wave
plasma etching. Porous SiCOH dielectric with a patterned TiN
hard mask was etched using the RF pulsed plasmas and the results
showed more anisotropic etching profiles with less sidewall
damages. Therefore, it is believed that the RF pulsed plasma
etching process of ultra low-k dielectric materials can improve the
RC delay related with plasma damage for the next interconnect
technology.

Introduction

In the development of advanced semiconductor integrated circuits, resistance and
capacitance (RC) time delays in copper interconnection became more critical than
complementary metal oxide semiconductor (CMOS) gate delay because the smaller pitch
of metal-to-metal makes the resistance and capacitance much higher. [1] In order to reduce
interconnect RC time delay, highly porous dielectric material, that is, porous SiCOH, is
used in recent interconnection to decrease the dielectric constant further. [2] However, the
porous SiCOH is very easily damaged during plasma etching processes by plasma
induced damages (PIDs). That is, during plasma etching process, plasma ions, reactive
radicals, and UV/VUV photons from the plasma break $Si\text{-}CH_3$ bonds in SiCOH and make
the material hydrophilic from hydrophobic.[3] As a result, the damaged porous SiCOH
easily bonds with moisture and degrades RC time delay by making dielectric constant
increased. Recently, many potential solutions to reduce PIDs during the integration with

porous SiCOH have been widely investigated using various methods such as pore stuffing[4-5], cryogenic etching[6-7], direct copper etching, etc.[8] However, they require additional processes, very low temperature, or a totally different integration scheme. Thus, BEOL(Back End of Line) integration with porous SiCOH still has challenges and it needs more researches to reduce PIDs.

In this study, RF pulsed plasma etching has been investigated to reduce the PID of porous SiCOH dielectric material. Compared to conventional plasma etching, RF pulsed plasma etching can vary etching characteristics such as selectivity, etch profile, PID, uniformity, UV/VUV radiation, and polymerization property by adjusting the pulse frequency and pulse duty cycle.[9-14] Among them, this study will specifically focus on the properties of RF pulsed plasma etching for less UV/VUV radiation, lower energy ion species, and higher polarization property to reduce plasma induced damage of an ultra low-k dielectric material. In copper/low-k interconnection, the weakest area to PID is the sidewall because the sidewall damaged layer remaining after the etching tends to increase the capacitance.[3] Thus, as a possible solution for the sidewall protection from the PID, an RF pulsed-plasma etching process which may have the properties of less UV/VUV radiation, lower energy ion species, and higher polymerization to the sidewall has been investigated. To prove this hypothesis, a dual-frequency capacitive coupled plasma system (DF-CCP) with a pulsed RF source (60 MHz) and a continuous wave RF bias (2 MHz) was used and the etching properties such as profile, the amount of PID, etc. by using pulsed plasma and conventional (CW) plasma were compared.

Experimental

<u>Equipment</u>

In this study, a dual frequency CCP system as shown in Figure 1 was used in the experiment. The top and bottom electrodes were covered by a silicon plate surrounded by a ceramic ring for the insulation. The distance between two electrodes for RF discharge was 80mm. The top electrode was perforated for gas flow to the chamber and a 60 MHz pulsed RF power source (Pearl CF-3000) was connected to control the plasma characteristics. A 2 MHz RF bias power (ENI GMW-50) was connected to the bottom electrode for ion bombardment onto the substrate and was cooled by water to maintain the substrate at room temperature. The chamber was evacuated by using a turbo molecular pump (3200 l/s) backed by a dry pump.

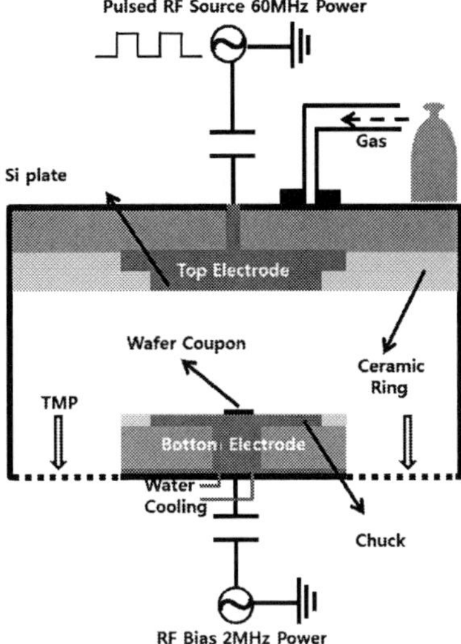

Figure 1. Schematic diagram of the dual frequency CCP reactor used in this study.

<u>Etch samples</u>

A thick SiO_2 and thin SiCN layer were deposited first on Si wafers for good adhesion between the porous SiCOH and the silicon wafer, and a porous SiCOH was deposited. On the SiCOH, a SiON/TiN/SiON multilayer hard mask pattern was formed for good etch selectivity. And the porous SiCOH was etched by using a DF-CCP system with a $C_4F_8/Ar/O_2/N_2$ gas mixture. The samples etched using both continuous wave plasmas and pulsed plasmas having 1 kHz frequency and 50% duty ratio were compared. The schematic diagrams of the sample for etching and RF pulsing waveforms are shown in Figure 2.

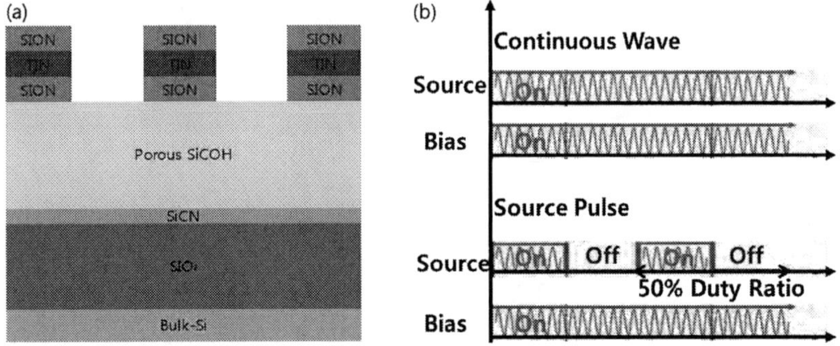

Figure 2. Schematic diagrams of (a) a porous SiCOH pattern structure and (b) continuous wave form and pulsed-wave form used in the experiment.

<u>Characterization</u>

The etch characteristics such as profile, etch depth, and etch rate of the porous SiCOH were estimated by Field Emission Scanning Electron Microscopy (FE-SEM, Hitachi S-4700). The plasma damage was characterized with FE-SEM by measuring sidewall thickness loss before and after dipping the etched samples in a diluted HF solution (0.5%, 60s). Figure 3 showed the schematic diagram of the etched profile before and after the dipping process for the estimation of the sidewall damage after the etching.

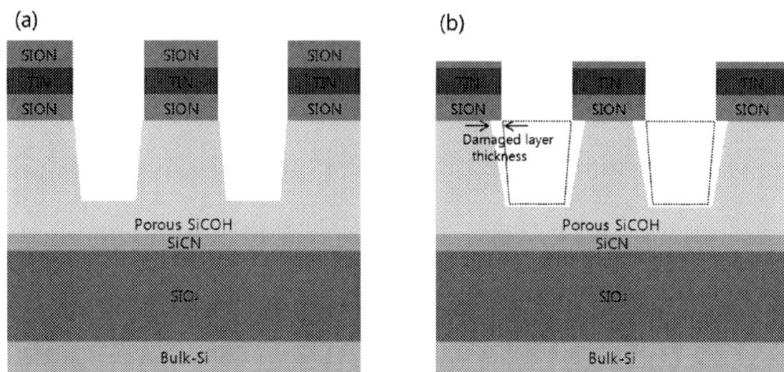

Figure 3. Schematic diagrams of etched profiles (a) before and (b) after a diluted HF dipping process.

Results and Discussion

<u>Profile Setup</u>

Before investigating the effect of RF pulsing, the etching by using conventional continuous wave (CW) plasmas was conducted for a proper etching condition. Figure 4 (a) – (c) show the SEM images of SiCOH profiles etched with various CW RF power conditions. In order to find a proper RF power, different combinations of source and bias power were conducted and the most anisotropic profile was obtained for the condition of Figure 4 (c), which was the condition of source 100W / bias - 600V.

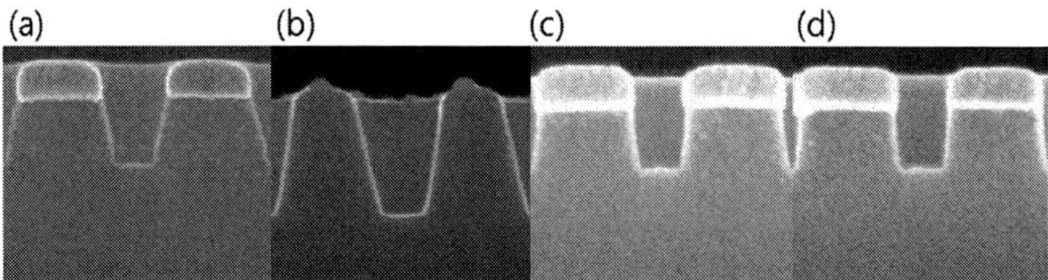

Figure 4. SEM images of etched profiles after the etching by (a) continuous wave / 200W source power / - 600V bias voltage, (b) continuous wave / 200W source power / - 1200V bias voltage, (c) continuous wave / 100W source power / - 600V bias voltage, (d) source pulse /100W source power / - 600V bias voltage.

Excessive source power (200W, Figure 4 (a)) made the profile less anisotropic and excessive bias voltage (-1200V, Figure 4 (b)) made the hard mask more eroded. This may be related to high energy ion species at higher powers and heavy ion bombardment for higher bias voltages. Using the plasma conditions in Figure 4 (c) by pulsing the source power at 50% and at 1kHz, the etched profile of SiCOH as shown in Figure 4 (d) was obtained. In general, compared to CW conditions, the etch profiles for the pulse conditions were more anisotropic. Detailed etch rates and sidewall slopes which indicates the criteria of anisotropic profile are showed in Figure 5.

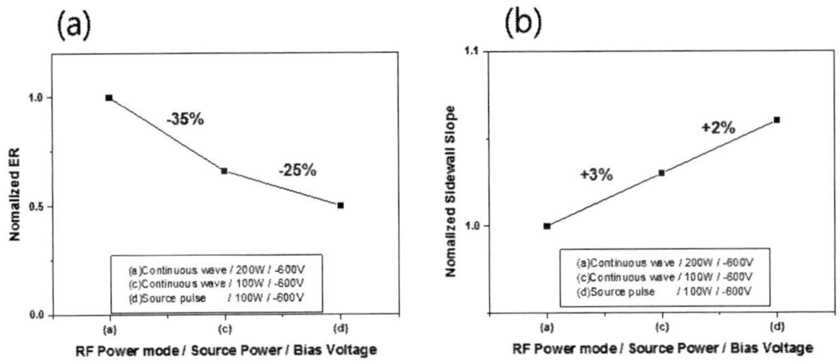

Figure 5. (a) Calculated etch rates and (b) measured sidewall slope angles for the etching of porous SiCOH etched patterns with various CW and pulse conditions.

In Figure 5, the data from Figure 4 (b) having a self-bias voltage of -1200V was removed because all mask layer was eroded. Figure 5 (a) shows the etch rates of porous SiCOH obtained from Figure 4 (a), (c), and (d). The etch rate for the condition of continuous wave / 200W source power / - 600V bias voltage was the highest possibly due to the high energy ion species and, after the source power was decreased by 100W, it dropped by 35%. On the contrary, when the pulse mode (50% at 1kHz) was turned on for the source power of 100W, the etch rate was just decreased by 25% from Figure 4 (c) condition. Figure 5 (b) shows the sidewall angle of the porous SiCOH patterns and, as shown in Figure 5 (b), the slope became more anisotropic as the source power was decreased and as the pulse mode was turned on. The most anisotropic etch profile was obtained for the etching condition with the source pulsing.

<u>Damage Characterization and Comparison</u>

In the previous section on profile setup, the optimum etching condition of CW plasma was the condition for Figure 4 (c) with the source power of 100W and the bias voltage of - 600V. In this section, the source pulse mode was turned on based on the optimum CW condition and plasma damage after the etching thick porous SiCOH was compared between two conditions.

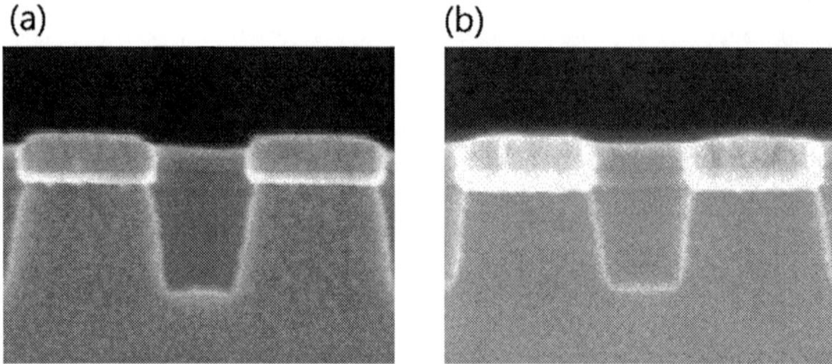

Figure 6. SEM images of SICOH etch profiles after dipping the etched sample in a diluted HF for 60s. (a) Profile for the CW condition (source power 100W / bias voltage - 600V) and (b) profile for the source pulse condition while keeping other conditions the same as (a).

Figure 6 showed the etch profiles after dipping the etched sample into diluted HF for 60s. As HF is used normally as a SiO_2 etchant, the plasma damaged layer on the porous SiCOH could be also etched away because the SiCOH surface is changed to Si-O-Si structure from Si-O-CH structure during the plasma etching process. [15] Both profiles showed the loss of a sidewall damaged layer, and the sidewall slope was degraded less anisotropic. However, Figure 6 (b) showed more anisotropic profiles and lower sidewall thickness loss than Figure 6 (a). Detailed damaged thickness loss ratio and slope change are shown in Figure 7.

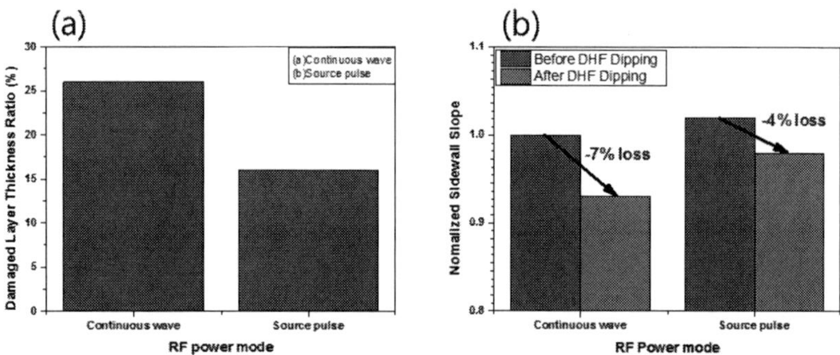

Figure 7. (a) Damaged layer thickness ratio per total CD and (b) sidewall slope change after dipping in a diluted HF solution.

In Figure 7 (a), the damaged layer thickness ratio per CD of the SiCOH for the pulse condition showed lower damage compared to that for the CW condition. Similarly, the sidewall slope change in Figure 7 (b) was less by 4% for the source pulse condition. These results indicate that the etching by using the RF source pulse mode damages porous SiCOH less than the etching by using the CW mode. This is because the pulsed plasma has characteristics of less UV/VUV radiation, lower energy ion species, and higher polymerization [12-14], all yielding lower damage compared to CW plasmas.[3] Also, the more anisotropic SiCOH etch profile obtained by the source pulse condition could result in not only less capacitance change but also less leakage current for the interconnection.

Conclusions

RF pulsed-plasma etching for porous SiCOH showed more anisotropic profile and less plasma damage characteristics compared to CW plasma etching due to less UV/VUV radiation and lower energy ionic species to the substrate. It also showed lower damaged layer thickness and less sidewall slope change after dipping in a diluted HF solution. Although CW plasma etching at an optimized condition of source power 100W / bias voltage - 600V showed a good etch profile, RF pulsed plasma etching showed a better etch profile and less damage. A disadvantage of using pulsed plasma compared to CW plasma seems to be the etch rate drop. However, it does not look significant because new IC nodes are also being scaled down. It is believed that RF pulsed-plasma etching can be applied as a low damage etching method in advanced IC generations which need more aggressive capacitance target.

References

1. R. H. Havemann and J.A. Hutchby, *Proc. IEEE*, **89**(6), 586-601 (2001).
2. A. Grill, S. M. Gates, T. E. Ryan, S. V. Nguyen, and D. Priyadarshini, *Applied Physics Reviews*, **1**, 011306 (2014).
3. M. R. Baklanov et al., *J. Appl. Phys.*, **113**, 041101 (2013)
4. T. Frot, W. Volksen, S. Purushothaman, R. Bruce, and G. Dubois, *Adv. Mater.*, **23**, 2828 (2011).
5. T. Frot, W. Volksen, S. Purushothaman, R. L. Bruce, T. Magbitang, D. C. Miller, V. R. Deline, and G. Dubois, *Adv. Funct. Mater.*, **22**, 3043 (2012).
6. L. Zhang, R. Ljazouli, P. Lefaucheux, T. Tillocher, R. Dussart, Y.A. Mankelevich, J.-F. de Marneffe, S. de Gendt, and M. R. Baklanov, *ECS J. Solid State Sci. Technol.*, **2**, N131 (2013).
7. M. R. Baklanov, J.-F de Marneffe, L. Zhang, I. Ciofi, and Z. Tokei, *Solid-State Technol.*, **57**, 5 (2014).
8. L. Wen, F. Yamashita, B. Tang, K. Croes, S. Tahara,T. Maeshiro, E. Nishimura, F. Lazzarino, I. Ciofi, J. Boemmels, and Z. Tokei, *IEEE IITC*, 173 (2015).
9. Samukawa S., *Appl. Phys. Lett.* **64**, 3398 (1994).
10. H. Ohtake and S. Samukawa, *Appl. Phys. Lett.*, **68**, 2416 (1996).
11. K. Tokashiki, et al., *Jpn. J. Appl. Phys.*, **48**, 08HD01 (2009).

12. M. Okigawa, Y. Ishikawa, and S. Samukawa, J. Vac. *Sci. Technol.*, B, **21**, 2448 (2003).
13. M. Okigawa, Y. Ishikawa, Y. Ichihashi, and S. Samukawa, *J. Vac. Sci. Technol.*, B **22**, 2818 (2004).
14. M. H. Jeon, K. C. Yang, K. N. Kim, G. Y. Yeom, *Vacuum* **121**, 294 (2015).
15. Q. T. Le, M. R. Baklanov, E. Kesters, A. Azioune, H. Struyf, W. Boullart, J.-J.Pireaux, and S. Vanhaelemeersch, *Electrochem. Solid-State Lett.*, **8**, F21 (2005).

ECS Transactions, 89 (3) 87-92 (2019)
10.1149/08903.0087ecst ©The Electrochemical Society

Capping Layer Effect on Lifetime of Plasma Etched Copper Lines

Mingqian Li, Jia Quan Su and Yue Kuo
Thin Film Nano & Microelectronics Research Laboratory
Texas A&M University, College Station, TX 77843, USA   yuekuo@tamu.edu

Copper is one of the most important interconnect materials in high-density ICs, large-area flat panel displays, and other electronic products. The new plasma-based etch method has been used in preparing copper fine patterns. Lifetimes of the plasma-etched copper lines with and without a TiW capping layer have been studied with the electromigration method. Under the same current density, the copper line with the TiW capping layer has a shorter lifetime than that without the capping layer. The diffusion of copper into the capping layer could enhance the void formation and therefore accelerate the failure of the copper line. The raise of the line temperature during electrical stress facilitated the broken of the line.

Introduction

Copper (Cu) is the most popular interconnect material for ICs, thin film transistor (TFT) arrays, etc. Compared with aluminum, Cu has a longer lifetime and larger conductivity. Kuo's group reported a room temperature plasma based Cu etch process (1-3). The Cu film was converted into the chloride or bromide compound at a high rate in a plasma reactor, which was subsequently dissolved in a HCl solution. This room-temperature process has been used in the fabrication of ICs and TFT LCDs (4). A modified method, i.e., replacing the HCl solution dip with a H_2 plasma exposure in an ICP reactor to remove the $CuCl_x$, has been reported (5-6). However, the H_2 plasma need to be carried at a low temperature, e.g., 10°C, and the $CuCl_x$ removal rate is low. In addition, the hydrogen plasma exposed photoresist is cross-linked and swollen (7), which is difficult to strip with a wet solution. A SiO_2 masking layer was required for that process.

An adhesion/barrier layer between the Cu line and the underneath substrate is usually required to increase the adhesion force and to prevent the Cu diffusion into the adjacent film. When a capping layer, e.g., Ag or SiN, is added on top of Cu, it can prevent the Cu diffusion to the subsequently deposited film as well as its oxidation in air (8-9). Electromigration (EM) is the phenomenon of the undesirable transfer of metal atoms during electrons moving under too high current stress conditions (10). It is commonly used to estimate the broken time of a metal line (11). The mechanical bending and geometry effects on the lifetime of the plasma etched Cu line have been studied using the EM method (12-13). In this paper, the TiW capping layer effect on the lifetime of the Cu line is investigated.

Experimental

Two types of samples, i.e., TiW barrier layer (10 nm)/Cu (280 nm) and TiW barrier layer (10 nm)/Cu (280 nm)/TiW capping layer (10 nm), were deposited on Corning glass. The TiW barrier and capping layers were sputter deposited under the same condition of 75 W, 5mTorr, Ar 50 sccm for 15 minutes. The Cu film was sputtered deposited at 80 W, 10 mTorr, Ar 30 sccm for 80 minutes. The TiW capping and barrier layers were plasma etched under the same condition of CF_4 10 sccm at 60 mTorr, 600W for 2 minutes. The exposed Cu was converted into $CuCl_x$ in the plasma of HCl/CF$_4$ 20/5 sccm at 70mTorr, 600 W for 2 minutes. Subsequently, the CuCl$_x$ was stripped off by dipping in an 8:1 diluted HCl solution. All plasma processes were done in a Plasma Therm 700C reactor under reactive ion etch (RIE) mode.

For the sample without the TiW capping layer, a non-uniform Cu chlorination reaction was observed. Since the Cu surface was exposed to air for a long time, e.g., more than 1 week, the formation of the native oxide might be the cause of the non-uniform reaction. Experiments on removing the native oxide on the Cu surface were carried out by dipping the sample in a 20% HCl solution before exposing to the HCl plasma. Cu surfaces before and after the HCl dip were analyzed with XPS to verify the above assumption. The EM test was carried out on the sample etched into the 4-point pattern as described in ref. 11. The sample was stressed under constant current density conditions at room temperature. The 'line broken' time when the resistance increases abruptly for several orders of magnitude was recorded as the failure time.

Results and Discussion

<u>Removal of native Cu oxide by HCl solution dip</u>

Figure 1 shows the Cu peaks in the XPS spectra of a Cu film surface (a) before and (b) after the HCl solution dip. The surface contains Cu in Cu$_2$O, CuO and metallic Cu forms. After the HCl dip, the intensity of the metallic Cu peak increases, the intensity of the Cu$_2$O peak reduces, and the intensity of the CuO remains almost the same. Table 1 lists the quantitative comparison of atomic ratios of the three components neglecting other surface elements, such as C and O, which are contributed by environmental contamination. After the HCl dip, the Cu concentration increased by about 17%, the Cu$_2$O concentration decreased by 12%, and the CuO concentration decreased by 4%. The Cu$_2$O/CuO ratio also increased after the HCl dip. Therefore, the native oxide, especially the partially oxidized Cu$_2$O, is effectively reduced by the HCl solution. It was observed that the removal of the native oxide greatly improved the Cu chlorination uniformity.

(a) 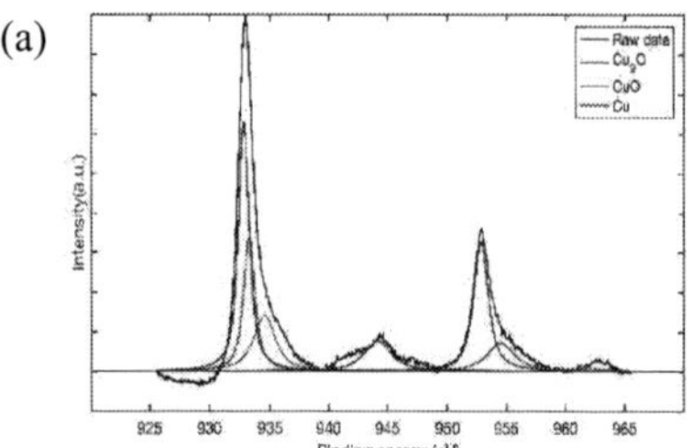 (b) 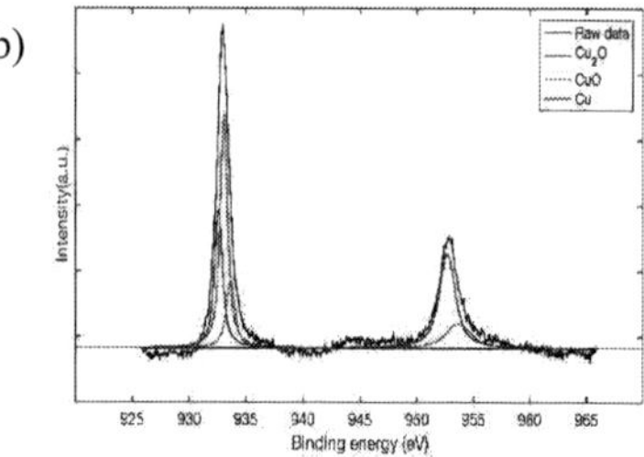

Fig. 1. XPS spectra of Cu surfaces (a) before and (b) after HCl solution dip.

Table. 1. XPS quantitative analysis of Cu components before and after HCl dip.

	Before dip	After dip
Cu/total Cu atomic ratio	0.25	0.43
Cu$_2$O/total Cu atomic ratio	0.60	0.47
CuO/total Cu atomic ratio	0.15	0.10
Cu$_2$O/CuO atomic ratio	4.03	4.49

<u>Capping layer effect on EM line broken time</u>

Figure 2(a) shows curves of the line broken time vs. stress current density of TiW capped and noncapped Cu lines of different widths. Samples were of the same length of 800 μm but different widths of 10 μm, 30 μm, or 2.5 μm. First, the line broken time reduced with the increase of the current density. This is consistent with the previous report that the lifetime decreased with the increase of the stress current density exponentially (14). Second, the 30 μm wide line has a shorter lifetime than the 10 μm wide line independent of the existence or absence of the TiW capping layer. This is also consistent with our previous observation that wide lines are subject to earlier failure than narrow lines due to the higher opportunity of forming voids at the larger number of grain boundaries (13). Third, based on the same line width, Cu with a TiW capping layer has a shorter lifetime than that without the capping layer. According to a previous study (15), the capping layer might induce void defects in the Cu layer, which could shorten the EM life time.

Figure 2(b) shows the same data as Fig. 1 expressed as the mean time to failure (MTTF) vs. current density. The following modified Black equation was used to fit each curve (10)

$$MTTF = \frac{A}{j^n} e^{\left(\frac{Q}{kT}\right)} = A \cdot e^{-n \cdot \ln j + \frac{Q}{kT}} = C \cdot e^{-n \cdot \ln j} \tag{1}$$

where A is a constant related to the cross section area, j is the current density, n is a model number related to the temperature, the stress condition and the conducting material, k is the Boltzmann's constant, Q is the activation energy, and T is the absolute temperature in K, and C is a constant dependent on A, Q and the absolute temperature T ($C = A \cdot e^{\frac{Q}{kT}}$).

In Fig. 2(b), the n-values of lines with same width are almost the same. The capping layer has no effect on the n-value. For the 30 μm and 2.5 μm wide lines, the n-values are 4.25-5.64. For the 10 μm wide lines, the n-values are 1.1-1.36. According to the previous study (16), the n-value is related to the EM test temperature. The Fig. 2(b) result indicates that lines of the same width were raised to the same temperature independent of the existence or absence of a TiW capping layer. For lines with the same width, the sample with the thin TiW capping layer should have a larger A-factor and therefore, a longer broken time. However, it was reported that the Black equation did not consider factors such as the temperature gradient and Joule heating effect (17). They might lead to the inaccurate n- and A-values.

Separately, it was reported that the temperature of the Cu line was increased by the EM stress (16-18). At high temperatures, Cu could easily diffuse into the TiW capping layer (19). Once Cu atoms are lost to the adjacent layer, voids could be formed in the bulk Cu line. In previous studies on the relationship between the Cu vacancy and diffusion (20-21), it was found that the increase of vacancy could enhance the Cu diffusion rate. Vacancies could even change the shape and location of the local grain boundaries (22). The following equation expresses the mass flux under the EM condition (14, 23)

$$J = \frac{N_A}{kT} \cdot D_0 \cdot e^{\left(-\frac{Q}{kT}\right)} \cdot (e \cdot Z^*) \cdot \rho \cdot j \tag{2}$$

where N_A is the density of lattice atoms, D_0 is the diffusion coefficient, Q is the activation energy, $e \cdot Z^*$ is the effective charge, ρ is the specific resistance, j is the current density, and k is the Boltzmann's constant. When the diffusion constant increases, the mass flux J increases. Then, the electromigration phenomenon becomes serious and the Cu line lifetime is shortened.

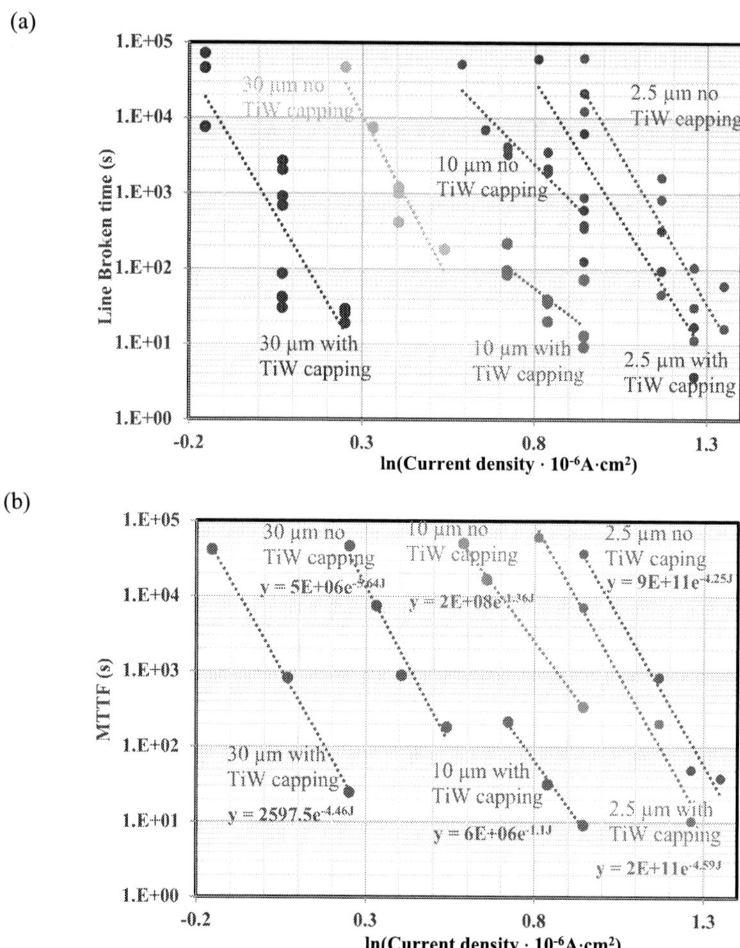

Fig. 2. (a) `Line broken` time vs. stress current density of TiW capped and uncapped Cu lines of different widths and (b) MTTF vs. current density of (a) samples fitted with Black equation.

The color of the TiW capped Cu line changed with the EM stress time. The original line color was purple, as shown in Fig. 3(a). It changed to the white color, as shown in Fig. 3(b), where the broken area showed as the dark color. Since the reflection of the light from the surface is dependent on the composition of the TiW capping layer, the color change may be related to its composition change. Table 2 shows the XPS elemental ratios of the TiW capping layer before and after the line broken. The W/Cu and Ti/Cu atomic ratios decreased after the EM stress while the Ti/W atomic ratio remained the same. Therefore, Cu was diffused into the TiW layer during the EM stress. The shortening of the line broken time from the addition of the TiW capping layer can be explained by the loss of Cu atoms in the bulk Cu line.

(a)

(b)

Figure. 3. Light microcopy images showing the color of a 30μm wide TiW/Cu/TiW line (a) before and (b) after line broken from EM stress at $J = 1.39×10^6 A/cm^2$.

Table. 2. XPS elemental ratios of TiW capping layer before and after EM stress

	Before	After
Ti/Cu atomic ratio	3.5	3.4
W/Cu atomic ratio	18.9	17.7
Ti/W atomic ratio	0.19	0.19

<u>Capping layer effect on Cu line resistance</u>

Figure 4(a) shows the resistances of 10 μm lines with and without TiW capping layer vs. the EM stress time at $J = 2.06×10^6 A/cm^2$. First, the resistance of the line changed little with increasing time until near the broken point. Second, the resistance of the sample with TiW capping layer is larger than that without the capping layer. According to previous studies (24-26), for Cu lines with the capping layer, EM prefers to happen at the interface. At the interface, Ti-Cu or W-Cu bonds provide lower conductivity than Cu-Cu bonds, which causes the larger resistance. Third, for lines without TiW capping layer, the broken phenomena occurred earlier than those without the TiW capping layer. This can be explained by the easy formation of voids in the bulk Cu line from the loss of atoms to the capping layer.

Figure 4(b) shows the resistance vs. EM stress time curves of the TiW capped and uncapped 2.5 μm lines at $J = 2.57×10^6 A/cm^2$. Similar to the 10 μm lines, the TiW capped lines had larger resistance than the uncapped lines. For both samples, the line resistance changed slightly with the increase of the stress time until near the broken point.

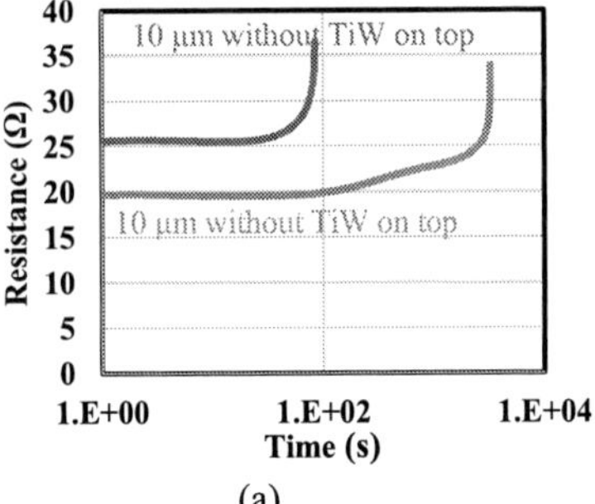

(a)

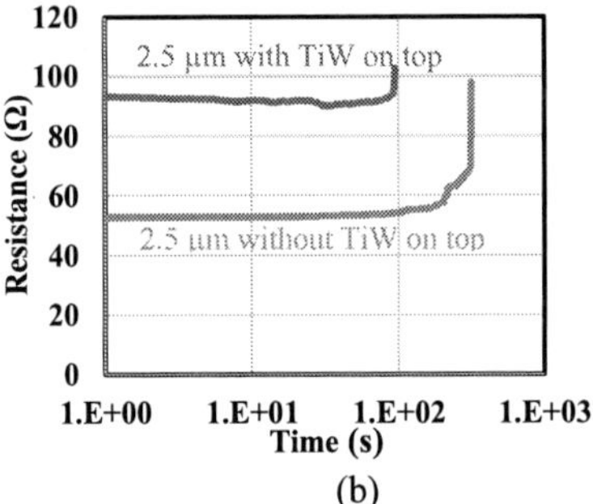

(b)

Fig. 4. Line resistance vs. stress time of TiW capped and uncapped Cu lines: (a) 10 μm line at $J = 2.06×10^6 A/cm^2$, (b) 2.5 μm line at $J = 2.57×10^6 A/cm^2$

Conclusion

The TiW capped Cu line has a shorter 'broken line' time than the uncapped Cu line. This can be explained by the loss of Cu atoms to the TiW layer during the EM stress. This phenomenon caused the enhancement of the vacancy formation and the acceleration of the line disruption. The raise of temperature during the current stress also enhanced the mass flux at the junction point that resulted in the early failure of the line. The resistance vs. stress time result showed that the TiW capping layer caused an increase of the line resistance leading to void formation in the bulk Cu line. Thus, the selection of the capping layer material is critical to the electromigration failure of the Cu line.

Acknowledgements

Authors acknowledge the financial support of this work through NSF CMMI project 1633580.

Reference

1. Y. Kuo and S. Lee, *Jpn. J. Appl. Phys.,* **39**(3AB), L188-L190 (2000).
2. Y. Kuo and S. Lee, *Appl. Phys. Lett.,* **78**(7), 1002-1004 (2001).
3. Y. Kuo and S. Lee, *Jpn. J. Appl. Phys.,* **41**(41), 7345-7352 (2014).
4. Y. Kuo, *Proc. 16th Intl. Workshop on Active-Matrix Flat Panel Displays and Devices,* 211-214 (2009).
5. P.A. Tamirisa, G. Levitin, N.S. Kulkarni, and D.W. Hess, *Microelectronic Engineering,* **84** 105-108 (2007).
6. F. Wu, G. Levitin and D. W. Hess, *ACS Appl. Mater. Interfaces,* **2**(8), 2175-2179 (2010).
7. Y. Kuo, *Jpn. J. Appl. Phys.,* **32**(1AB), L126-L128 (1993).
8. M. C. Kang, Y. J. Kim and J. J. Kim, *Electrochemical and Solid-State Letters,* **12**(9), H340-H343 (2009).
9. Y. Kuo, *J. Electrochem. Soc.,* **137**(4), 1235 (1990).
10. J. R. Black, *IEEE Transactions on Electron Devices,* **16**(4), 338-347 (1969).
11. J. R. Black, *Reliability Physics Symposium IEEE,* 142-149 (1974).
12. G. Liu and Y. Kuo, *J. Electrochem. Soc.,* **156**(7), H579-H584 (2009).
13. M. Li and Y. Kuo, *ECS. Trans.,* **86**(8), 41-47 (2018).
14. A. Buerke, H. Wendrock and K. Wetzig, *Crystal Res. Technol.,* **35**(6-7), 721-730 (2000).
15. X. Lu, J. W. Pyun, B. Li, N. Henis, K. Neuman, K. Pfeifer and P. S. Ho, *Proceedings of the IEEE International,* 33-35 (2005).
16. G. Liu, *Dissertations & Theses - Gradworks* (2008).
17. M. Shatzkes and J. R. Lloyd, *J. Appl. Phys.,* **59**(11), 3890-3893 (1986).
18. A. H. Fischer, A. V. Glasow, S. Penka, and F. Ungar, *IEEE International Interconnect Technology Conference,* 139-141 (2002).
19. C. Ryu, H. Lee, K. W. Kwon, A. L. Loke and S. S. Wong, *Solid State Technology,* **42**(4), 53-56 (1999).
20. M. R. Sørensen, Y. Mishin and A. F. Voter, *Physical Review B,* **62**(6), 3658 (2000).
21. A. Suzuki and M. Yu, *Interface Science,* **11**(1), 131-148 (2003).
22. V. Vitek, Y. Minonishi and G.J. Wang, *J. de Physique,* **46**, 171 (1985).
23. E. Arzt, O. Kraft and R. Spolenak, *Z. Metallkunde,* **87**, 934-942 (1996).
24. M. W. Lane, E. G. Liniger and J. R. Lloyd, *J. Appl. Phys.,* **93**(3), 1417-1421 (2003).
25. J. R. Lloyd and J. J. Clement, *Thin Solid Films,* **262**, 135 (1995).
26. K. Hu, R. Rosenberg and K. L. Lee, *Appl. Phys. Lett.,* **74**, 2945 (1999).

Chapter 6

Technologies for Advanced Integrated Circuits

ECS Transactions, 89 (3) 95-110 (2019)
10.1149/08903.0095ecst ©The Electrochemical Society

Cavity-Free Micro Thermoelectric Energy Harvester with Si Nanowires

T. Watanabe[a,b], M. Tomita[b], T. Zhan[b], K. Shima[a], Y. Himeda[a], R. Yamato [a],
T. Matsukawa[c], and T. Matsuki[b,c]

[a] Faculty of Science and Engineering, Waseda University,
Shinjuku, Tokyo, 169-8555, JAPAN
[b] Research Institute for Ambientornics, Waseda University,
Shinjuku, Tokyo, 169-8555, JAPAN
[c] National Institute of Advanced Industrial Science and Technology,
Tsukuba, Ibaraki, 305-8568, JAPAN

We present a new design of silicon-based micro-thermoelectric
generator, which utilizes silicon nanowires as the thermoelectric
leg. It is driven by a steep temperature gradient exuding around a
heat flow perpendicular to the substrate, and the silicon nanowires
are not suspended on a cavity etched on the substrate. The power
density is scalable by shortening the silicon nanowire to sub-μm
length, which was experimentally demonstrated and tens of
μW/cm^2-class power generation was achieved at an externally
applied temperature difference of only 5 K. A numerical discussion
shows that the thermoelectric power can be drastically enhanced by
suppressing the thermal resistance at the entire substrate. Thus,
there is a plenty of room at the micro- or submicrometric scales for
realizing thermal energy harvesting devices with high power
densities.

Introduction

Realization of the trillion-sensors-network society is owing to the maturity of energy
harvesting technology which generates electric power from environmental energy sources
such as light, heat, vibration, motion, or ambient RF. Among various energy harvesting
devices, the thermoelectric generator (TEG), which utilizes the Seebeck effect of
semiconductors, is anticipated as an ultimate one because it has a potential to provide a
semi-permanent power from almost unlimited heat energies.

In a TEG, the efficiency of energy conversion from heat to electricity is determined
by the non-dimensional thermoelectric figure of merit, $ZT = \alpha^2 \sigma T/\kappa$ (1) of the employed
semiconductor, where α [V/K] is the Seebeck coefficient, σ [Ω^{-1}cm^{-1}] the electrical
conductivity, κ [W cm^{-1} K^{-1}] the thermal conductivity, and T [K] the average temperature
between heat source and cold sink. Although a higher ZT is favorable to achieve high
conversion efficiency, only limited materials were known to have high ZT near room
temperature, such as (Bi, Sb)$_2$(Se, Te)$_3$ alloys.

Recently, silicon nanowires (Si-NWs) emerged as a promising thermoelectric
material; the thermal conductivity of Si-NW was found to decrease drastically as the
nanowire is thinned (2,3). Since then, a number of fabrication examples of Si-based TEGs
has been reported (4), including planar type TEGs (5-7) and vertical types using
perpendicularly oriented Si-NW legs (8-14). As shown in Fig 1, conventional
demonstrations of planar TEGs employed Si-NWs about 10-100 μm long which were

suspended on a cavity to avoid the bypass of the heat current. Fabrication of the vertical TEGs requires a deep substrate etching process or controlled growth of vertical Si-NWs.

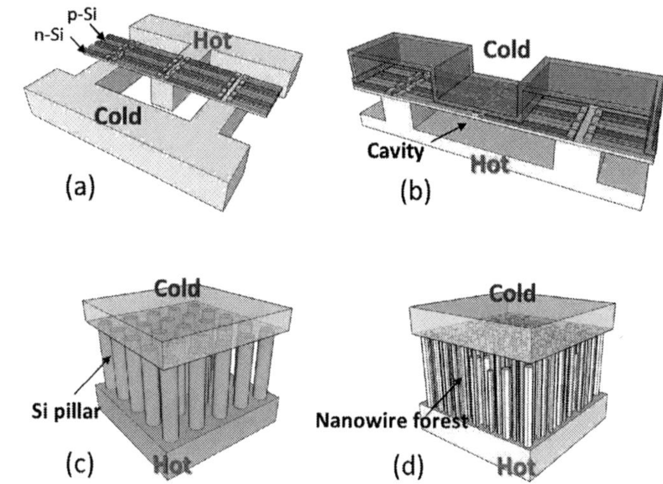

Figure 1. Schematics of Si-based micro-thermoelectric generator devices. (a) Planar type TEG suspended between thermally separated substrates (5). (b) Planar type TEG suspended on a cavity on the substrate (6, 7). The external temperature difference is applied between top and bottom surfaces. (c) Vertical type TEG with Si pillar legs fabricated by a deep reactive ion etching process (8,9). (d) Vertical type TEG using randomly grown/etched Si-NW nanowire forest (10-14).

The authors' group proposed a cavity-free design of planar and short Si-NW TEG (15-19). The conventional TEGs were designed to ensure the conversion efficiency by making the leg part as long as possible, thereby the parasitic heat resistance can be relatively suppressed. Contrary to the conventional design philosophy, our proposed device is designed by focusing on the enhancement of the output power density. The leg part is kept rather shorter to obtain a larger output power by integrating many TEGs in series. The TEG is operated by a steep temperature gradient exuding around a heat flow perpendicular to the substrate, and the leg is no need to be suspended on a cavity.

In this paper, the design concept of the new TEG device is presented in detail. An experimental demonstration of test devices (18,19) is reviewed. The possibility of further improvement in the output power is discussed by focusing on the parasitic heat resistance at the substrate.

New Device Architecture

Before showing the new device structure, we explain how we figured out it. Figure 2 shows examples of the temperature distribution of planar Si-NW TEG structures simulated by finite-element-method (FEM) (20). Fig. 2a shows a conventional planar device in which the Si-NWs and supporting insulating SiO_2 membrane are suspended on a cavity etched in the Si substrate. The Si-NWs are located between the comb-shaped thermal conductivity pads made of aluminum nitride (AlN). A uniform temperature gradient formed in the Si-NW part. Fig. 2b shows a similar device structure but it differs in the substrate details. Here, Si-NWs are not suspended on a cavity, and the temperature gradient in the Si-NW region almost vanishes. However, we realized a very steep temperature gradient that remained in the very vicinity of the heat source. The temperature slope ranges about 200 nm. Therefore, if the length of Si-NW leg is downscaled at submicrometer scale, a steep temperature gradient can be easily applied to them without forming a cavity structure.

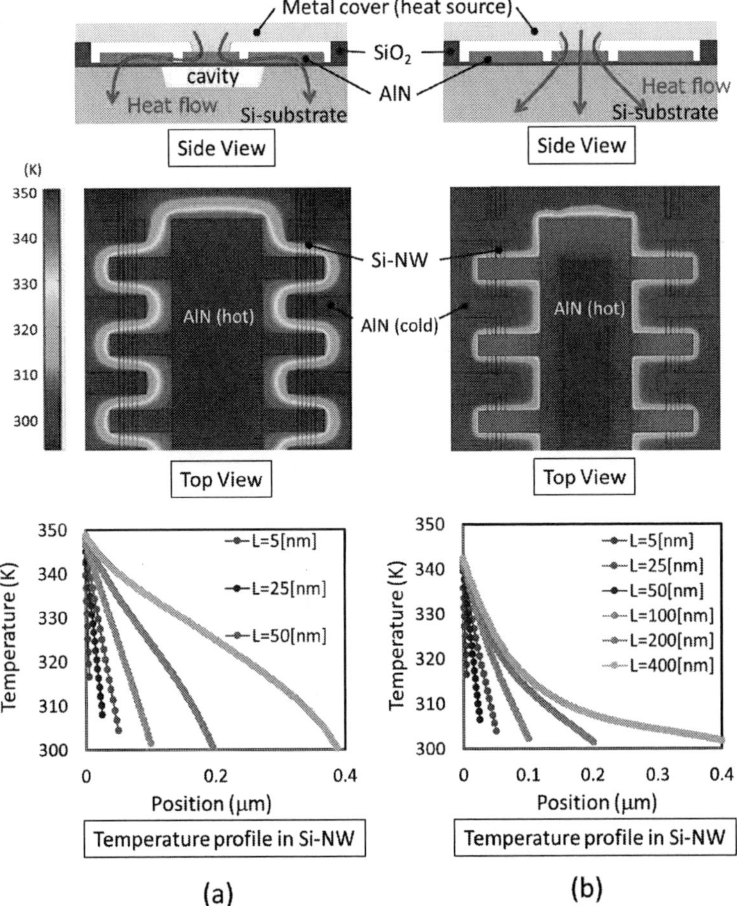

Figure 2. Simulated temperature distributions of two types of planar Si-NW TEGs (20). (a) Conventional planar device in which the Si-NWs and supporting SiO_2 membrane are suspended on a cavity etched in the Si substrate. (b) Planar device without cavity in the Si substrate.

Figure 3 shows the schematics of the newly proposed TEG structure. The Si-NW legs are placed on the substrate without forming a cavity space underneath the buried oxide (BOX) layer. The heat current flows perpendicularly to the substrate, and the TEG is driven by the steep temperature gradient exuding around the heat flow. Since the temperature gradient is concentrated in the limited range of a few hundred nm, it is effective to make the leg length shorter than or equal to the slope region. The output voltage is elevated by alternatively connecting the n-type and p-type legs in series. By injecting the heat to the substrate with a fine pitch, a large number of legs can be designed to cover the whole surface and thus the areal power generation density can be enhanced. In other word, the power density is scalable by shrinking the Si-NW length and heat injection pitch.

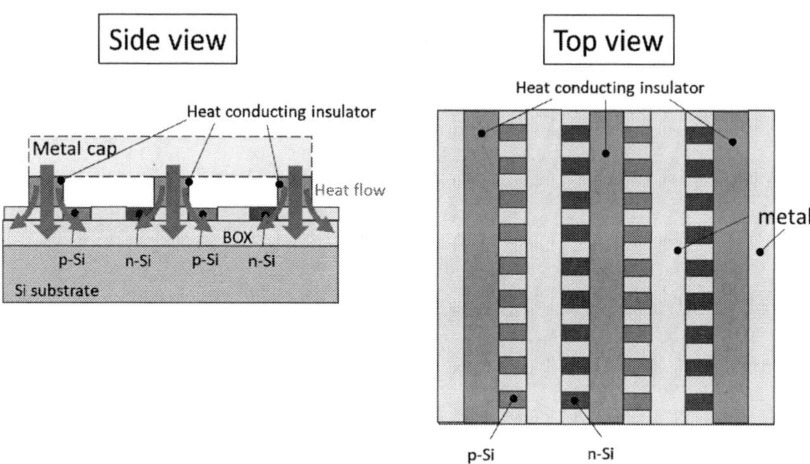

Figure 3. Schematic of the newly proposed planar-type Si-NW TEG structure (15-19). The TEG is driven by a steep temperature gradient exuding around a heat flow perpendicular to the substrate. Si-NWs are not suspended on a cavity etched in the substrate.

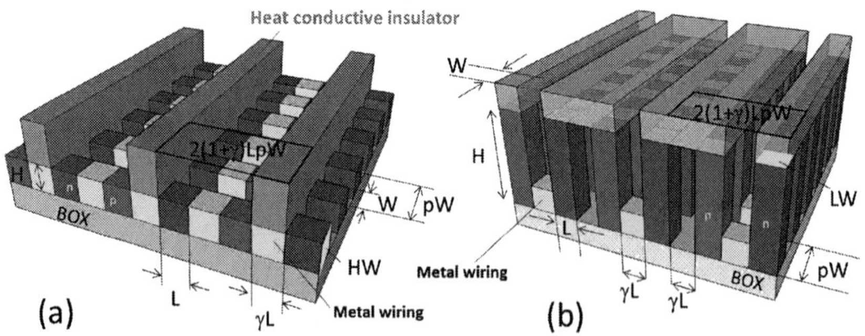

Figure 4. Dimensions of a TEG to analyze the scaling rule of the areal power density (15, 16). (a) The newly proposed planar-type Si-NW TEG. (b) Conventional π-shaped TEG.

The scalability of the power density is the most important feature of the proposed TEG architecture. Figure 4 shows the scaling rules of the areal power density for the proposed planar TEG (Fig. 4a) and that for the conventional π-shaped TEG structure (Fig. 4b). In the planar type TEG, the maximum value of the areal power density is given by

$$P_{max,planar} = \frac{\alpha^2 (HW)\sigma\Delta T^2}{2L} \frac{1}{2(1+\gamma)LpW} = \frac{\alpha^2 H\sigma\Delta T^2}{4(1+\gamma)pL^2}, \qquad [1]$$

where H, L, W are the height (vertical thickness), length and width of the thermoelectric leg, respectively. ΔT is the temperature difference across the single leg. γ is the ratio of the wiring to the leg length, thus γL is the length of the wiring metal part. p (>1) is the ratio of the leg pitch to the leg width, thus pW corresponds to the leg pitch. The areal power density of the planar type TEG is inversely proportional to L^2, so that the thermoelectric performance is improved by shrinking the nanowire length.

Contrary, the power density of the conventional π-shaped TEG is given by

$$P_{max,vertical} = \frac{\alpha^2 (LW)\sigma\Delta T^2}{2H} \frac{1}{2(1+\gamma)LpW} = \frac{\alpha^2 \sigma\Delta T^2}{4(1+\gamma)pH}. \qquad [2]$$

In this case, L and W are the width of vertical leg, and H corresponds to the leg length, as shown in Fig. 4b. The areal power density is independent of L and W, thus it does not benefit from lateral scaling.

Numerical Analysis

In the above discussion, ΔT is treated as an independent parameter, but it must depend on the dimensions of the leg. In what follows, we provide a more accurate analysis on the scaling rule of the proposed planar TEG (16,18,19) by means of the FEM using COMSOL Multiphysics. The TEG performance was simulated by simultaneously solving the heat conduction equation, the Poisson equation, the charge continuity equation, the drift-diffusion current equation, the Seebeck equation, and the Peltier equations. Parameter values can be found elsewhere (16,19).

Figure 5 shows the simulated TEG model, which is based on a unit structure of a 2-dimensional periodic boundary condition. The temperature difference between the top heater and the cold stage is set to 5 K. Si-NWs are laid on a BOX layer with 145 nm thick. The thickness H and width W of the Si-NW were 50 and 100 nm, respectively. Both ends of the Si-NW were connected to wiring electrodes of Al. The length of the wiring electrodes is set to be same as the Si-NW length L, namely $\gamma = 1$. The open space is assumed to be filled with air. The electrical resistivity values σ of n-type and p-type Si-NWs are set to 4.3×10^{-3} Ω cm. The Seebeck coefficient α is ± 200 μV/K.

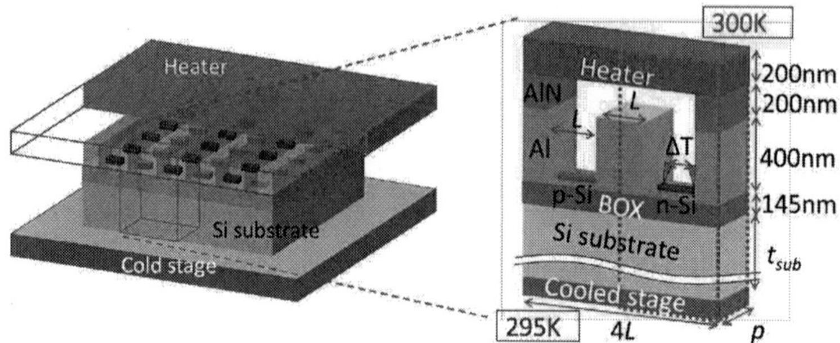

Figure 5. Simulation model of planar Si-NW TEG without cavity in the Si substrate (18,19). Left is an entire structure including heater and cold stage. Right is the unit cell structure under the two-dimensional periodic boundary condition.

Figure 6 shows the simulation result of the temperature difference ΔT across the Si-NW and the areal power density at a Si substrate thickness of 745 µm. Although the ΔT decreases as the Si-NW shortens, the power density still keeps on increasing in the range of $L > 0.1$ µm.

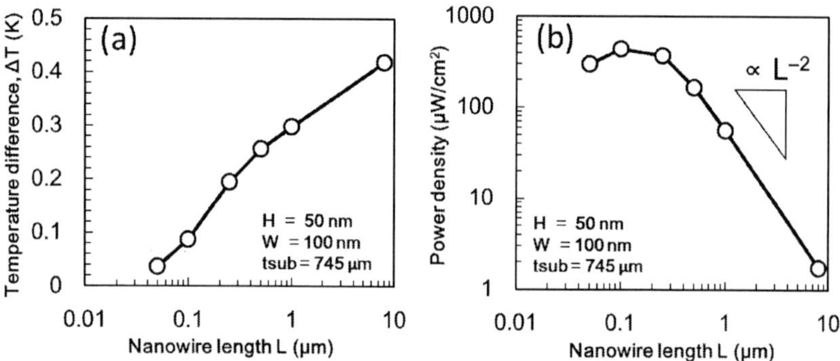

Figure 6. Results of TEG device simulation at an externally applied temperature difference of 5 K. (a) Temperature difference across the Si-NW. (b) Areal density of thermoelectric power (18,19).

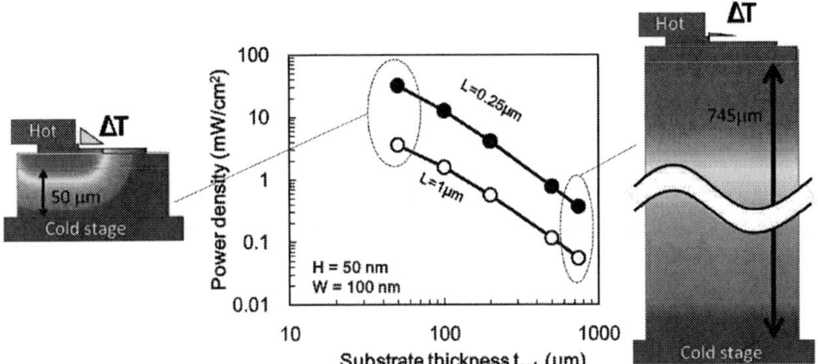

Figure 7. Si substrate thickness dependency of the power density simulated by a FEM device simulation at an externally applied temperature difference of 5 K (18,19).

The temperature difference ΔT across the Si-NW is much smaller than the externally applied temperature difference of 5 K. This means that most of the temperature difference is applied to the other part which is acting as a parasitic heat resistance over the resistance at Si-NW leg. The largest one may be for a Si substrate with 745 µm thick. Therefore, the power density is expected to be improved by employing thinner substrates.

Figure 7 depicts the simulated the power density as a function of the Si substrate thickness at the externally applied temperature difference of 5 K. Thus, substrate thinning is expected to have a considerable effect on the power density.

Experimental Demonstration

Experimental Method

It has been experimentally confirmed that the thermoelectric power was enhanced by shortening the NW (17) and thinning the Si substrate (18,19). Figure 8 shows the fabrication process of a test TEG device on a silicon-on-insulator (SOI) substrate using conventional CMOS processing. The SOI layer was patterned into Si-NWs by ArF immersion lithography and reactive ion etching. The SOI layer was doped with phosphorus ions with a dose of 5.0×10^{15}cm^{-2}. The width of Si-NW is about 65 nm, and the length is varied from 1.0 µm to 0.25 µm. Both ends of Si-NW bundle are connected to Si-pads, on which Al/TiN/Ti electrode was deposited. One side of the electrode was heated by attaching a custom-made micro-thermostat. The micro-thermostat comprises AlN ceramic plate on which a resistive heater and a platinum temperature sensor to control the temperature of the AlN plate. The electric contact of the hot-side electrode was prepared separately, which is placed next to the cold-side electrode and connected to the hot-side electrode via a narrow leader line. The thickness of the base Si-substrate was varied from 745 µm to 50 µm by thinning with backside grinding.

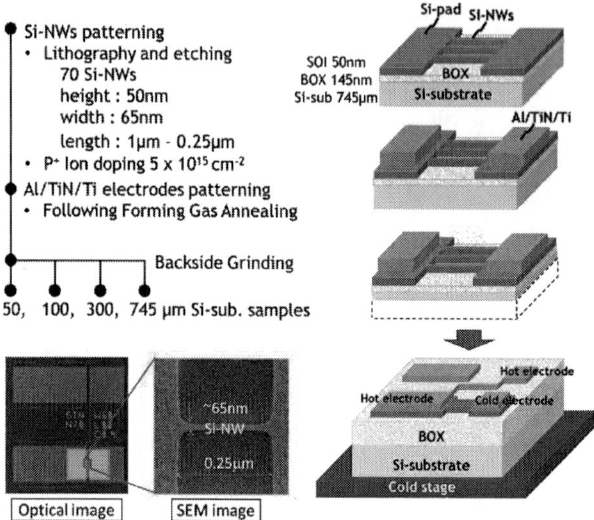

Figure 8. Fabrication process flow and as-built images of the test TEG device structure (18,19).

Figure 9 shows the measurement set-up for the thermoelectric characteristics. The micro-thermostat is used as the heat source, which is kept at about 297 K. The sample stage acts as a heat sink, whose temperature was kept at 292 K by a water chiller and a Peltier controller. The thermoelectric performance was evaluated by the usual current–voltage (I–V) characteristic measurement. The thermoelectric generator is merely a resistive element under no application of the temperature difference, so that the I-V relation shows a linear ohmic characteristics. By applying a temperature difference to the device, the linear I–V curve shifts away from the origin, and the product of the measured current and loading voltage corresponds to the thermoelectric power. Varying the loading voltage corresponds to varying the load resistance.

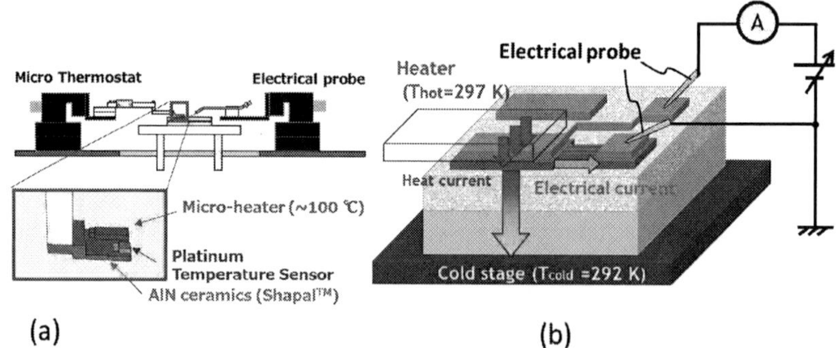

Figure 9. Schematics of the experimental setup for the thermoelectric characteristics measurement (18,19). (a) Electrical probe system equipped with a micro-thermostat. (b) Set-up for the TEG device operation.

<u>Experimental Results</u>

Figure 10 shows the measured thermoelectric power of the fabricated Si-NW devices. Fig. 10a shows the power curves as a function of the loading voltage, measured under the externally applied temperature difference of 5 K. Figure 10b shows the maximum power density and the internal electric resistance, as a function of the Si-NW length. The power density is given by normalizing the measured power with the area of the product of twice of the NW length and the NW pitch. This normalization is assuming that the pads and NW arrays have a same footprint in an integrated module of the planar TEG. The maximum power is obtained in the shortest NW device with 0.25 μm length.

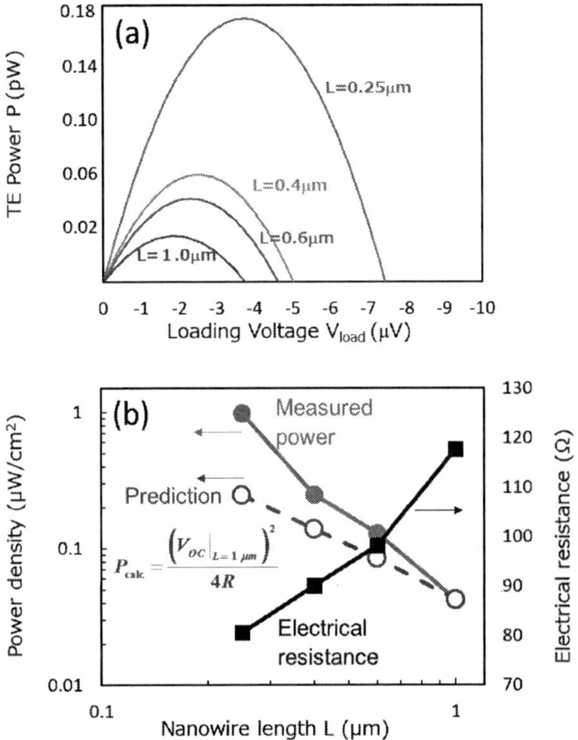

Figure 10. Thermoelectric power of fabricated Si-NW TEGs measured at an externally applied temperature difference of 5 K (18,19). (a) Power – loading voltage curves. The maximum power and the open circuit voltage V_{oc} increase as the Si-NW length shortens. (b) The maximum power density and the internal electric resistance as a function of the Si-NW length L. P_{calc} is a simple estimation by assuming a constant V_{oc} obtained for L= 1.0 μm. The experimentally measured power values are much larger than those obtained from a simple estimate.

It is not a surprising result that the maximum power P_{max} is enhanced by shortening the Si-NWs, because it is given by

$$P_{max} = \frac{V_{OC}^2}{4R},\qquad\qquad [3]$$

where V_{oc} is the open circuit voltage and R is the internal resistance of the thermoelectric generator. P_{max} should be increased in the shorter leg devices with smaller resistance R. However, the experimental result exhibits much higher power output than the simple prediction. In Fig. 10b, predicted maximum power P_{calc} is also shown, which is calculated by assuming a constant V_{oc} obtained at the Si-NW length of 1.0 μm. The experimentally measured power is much larger than the simple prediction.

In Fig. 10a, the open circuit voltage V_{oc} also increases as the Si-NW length shortens. This is an unexpected result, considering that the open circuit voltage is given by $V_{oc}=\alpha\Delta T$. The temperature difference ΔT across a Si-NW must decrease as the Si-NW shortens as shown in Fig. 6a. Thus V_{oc} was expected to decrease with downscaling the device dimension. Our experimental result shows an opposite behavior, suggesting that the Seebeck coefficient α is enhanced effectively by shortening the Si-NWs. The increase of V_{oc} in shorter NWs was observed repeatedly in our previous experiments (15, 17).

The physical origin of the Seebeck coefficient enhancement remains an open question. We have considered two possibilities: enhanced phonon drag effect (21) and the appearance of the Benedicks effect (22) in shorter Si-NWs, but none of these models are convincing. This topic will be discussed later in detail.

The thermoelectric power was drastically enhanced by thinning the Si-substrate. Figure 11 shows the dependence of the power density on the Si-substrate thickness. The power density of a sample with 50 μm thick Si-substrate reached 12 μW/cm². The enhancement can be attributed to the increased temperature difference across the Si-NWs by suppression of the series thermal resistance at the Si-substrate.

The 10 μW/cm²-class generator may open up the way for various IoT applications (19). The highest power consuming levels are in the application of wireless transmitters, which require a few microwatts of power even in sleep mode. Autonomous and perpetual operation of small sensor nodes can be realized by suppressing the transmission data amount to a few Mbit information once every hour (19).

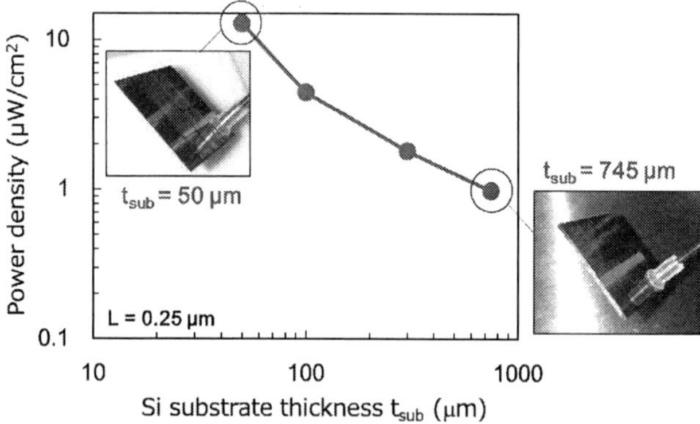

Figure 11. Dependence of the measured power density on the Si-substrate thickness. (18,19)

Discussion

<u>Origin of the Seebeck coefficient enhancement in short Si-NWs</u>
As shown in Fig. 10, it was found that the Seebeck coefficient of the Si-NWs increased upon decreasing their length. Figure 12 shows two possible mechanisms: (a) an enhanced phonon drag effect (21), and/or (b) the appearance of the Benedicks effect (22). However, these mechanisms are not yet convincing, and the physical origin remains an open question.

Fig. 12a schematically depicts the behavior of phonons in the Si-NWs. The phonon drag effect is that the momentum of a phonon flux is transferred to the electron gas leading to an enhancement of the Seebeck coefficient. It is primarily a low temperature effect, but its contribution is non-negligible in crystalline Si even at room temperature (21). In long Si-NWs, the phonon drag effect will be diminished by frequent surface scattering. Contrary, the phonon drag in short Si-NWs is expected to be effective thanks to the low surface scattering frequency, and thus the Seebeck coefficient of the bulk silicon comes back again to some extent. The scenario seems reasonable, but it is inconsistent with the typical feature of ballistic transport phenomena; the conductance is saturated in the short range and thus the conductivity rather decreases as the channel length decreases.
The other possible mechanism, the appearance of the Benedicks effect is illustrated in Fig. 12b. The Benedicks effect is a thermoelectric phenomenon in which the themoelectric power depends on the temperature gradient. In the well-known Seebeck effect, the thermoelectric power depends only on the temperature difference across a thermoelectric material, so that it is independent of the length. As shown in Fig. 12b, the Benedicks effect develops a potential between two points in a conductor at the same temperature but separated by a non-zero temperature gradient. The Benedicks effect is much smaller than the Seebeck effect.

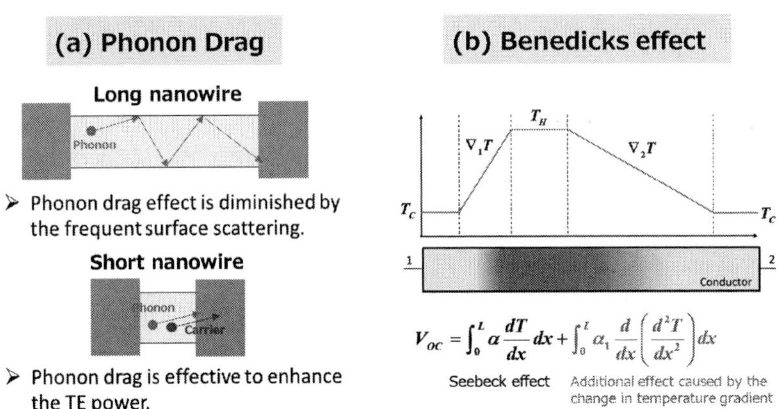

$$V_{OC} = \int_0^L \alpha \frac{dT}{dx}\,dx + \int_0^L \alpha_1 \frac{d}{dx}\left(\frac{d^2 T}{dx^2}\right) dx$$

Figure 12. Possible mechanisms to explain the Seebeck coefficient enhancement in short Si-NWs: (a) enhanced phonon drag effect (18, 19, 21), and (b) appearance of the Benedicks effect (22). However, these mechanisms are not yet convincing as discussed in the text.

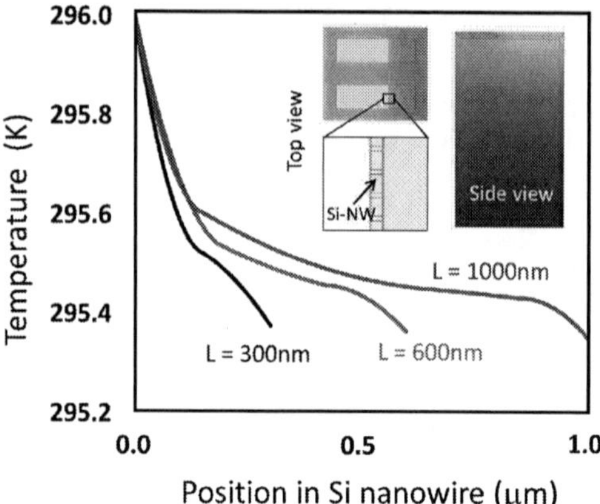

Figure 13. Simulated temperature profiles in Si-NWs of the test planar TEG.

Under a significantly large temperature gradient, the thermoelectric open circuit voltage can be written as

$$V_{OC} = -\int_0^L \alpha \frac{dT}{dx}dx - \int_0^L \alpha_1 \frac{d}{dx}\left(\frac{d^2T}{dx^2}\right)dx + \cdots, \tag{4}$$

where α[V/K] is the conventional Seebeck coefficient, and α_1 [Vm^2/K] is the coefficient of the lowest-order nonlinear term, corresponding to the Benedicks effect. Grigorenko *et al.* (22) formulated the coefficient α_1 based on the Boltzmann transport equation as

$$\alpha_1 = \frac{14\pi(k_B\tau)^2}{5me}T. \tag{5}$$

Here, τ is the scattering life time of carries, and m is the effective mass of the carrier, and e is the elementary charge. In our proposed planar device structure, the temperature distribution is non-linear as shown in Figure 13. A steep temperature gradient is established in the vicinity of the hot-side electrode. The temperature profile in the Si-NW can be approximated by

$$T(x) = T_{hot} - \frac{\Delta T}{1-\exp(-L/l)}\left(1-\exp\frac{-x}{l}\right), \tag{6}$$

where x is the longitudinal position component in the Si-NW, and l is the characteristic length of the temperature decay. By substituting T for Eqn. 6 in Eqns. 4 and 5, we get

$$V_{OC} = \alpha\Delta T + \frac{14\pi(k_B\tau)^2}{5me}\left\{\frac{T_{hot}\Delta T}{l^2} - \frac{\Delta T^2}{2l^2}\right\} + \cdots, \tag{7}$$

$$\alpha_{eff} = \alpha + \frac{14\pi(k_B\tau)^2}{5mel^2}\left\{\frac{T_{hot}+T_{cold}}{2}\right\} + \cdots. \tag{8}$$

The second terms on the right-hand sides of Eqns. 7 and 8 correspond to the contribution from a non-linear large temperature gradient.

According to the FEM simulation shown in Fig. 13, the decay length l is about 200~300 nm. Assuming that $\tau = 1.0 \times 10^{-12}$ s and $m = 0.4 m_0$, the second terms in the right-hand side of Eqn. 8 is estimated to be 0.4~0.7 µV/K. These values are too small to explain the enhancement in the effective Seebeck coefficient observed in the experiment. The assumed scattering lifetime τ should be decreased in the Si-NW, so that the contribution from the non-linear large temperature gradient will further decrease. Thus, it is difficult to explain the enhanced thermoelectric power in short Si-NWs by only the Benedicks effect.

The other possible origins may be hidden outside of the Si-NW, e.g., the electrode pad and its junction part to the Si-NW. Although we confirmed experimentally that the thermoelectric power generated in the one electrode is negligibly small, the contribution from such series components may be a non-negligible origin to modulate the effective Seebeck coefficient of the whole device.

Possibility of the Si-substrate Engineering

As shown in Fig. 11, the thermoelectric power was drastically enhanced by thinning the Si-substrate. Therefore, further improvement of thermoelectric power can be expected by suppressing the thermal resistance at the substrate. For example, the temperature difference across the Si-NW is expected to increase by partially replacing the Si substrate underneath the cold side edge of the Si NW by Cu which has a higher thermal conductivity.

We pursued an optimum substrate structure to create a large temperature difference across the Si-NWs, by means of FEM simulation (24). Figure 14 shows the cross-sectional view of the unit structure of the planar Si-NW thermoelectric generator. The substrate part is divided into three regions, Sub.1, Sub.2, and Sub.3. Various substrate structures were tested by changing the material of each region. The substrate structure is described as the materials of Sub.1-Sub.2-Sub.3. In this work, we prepared Si-Si-Si, Si-Si-Cu, Cu-Si-Si, Ox-Si-Cu, Cu-Si-Ox, Cu-Si-Cu and Cu-Cu-Cu substrates, where SiO_2 is expressed as Ox.
The thicknesses of the Al, AlN, NiSi, Si-NWs, SiO_2 layers are set to 100 nm, 550 nm, 50 nm, 50 nm, and 145 nm, respectively. The width and pitch of the Si-NWs are 100 nm and 400 nm, respectively. The length of the Si-NWs and the NiSi pad between them is fixed to 200 nm. The thermal conductivities of Cu, Si, SiO_2 and SiNWs are set to 400 W/mK, 131 W/mK, 1.4 W/mK and 1.4 W/mK, respectively. The heat transport calculation is based on the Fourier rule.

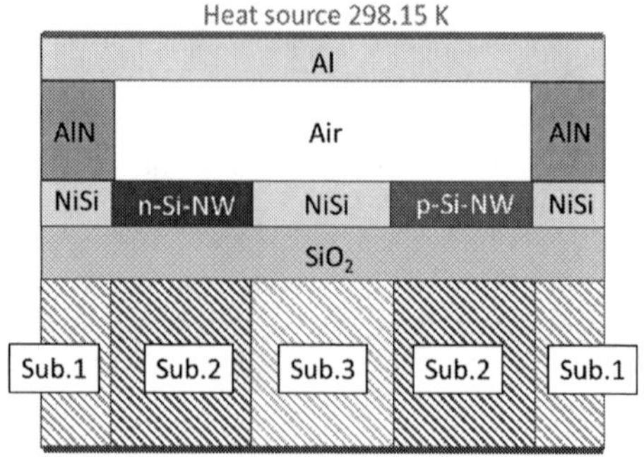

Figure 14. Simulation model of the planar-type TEG on the substrate with spatially varying material (24).

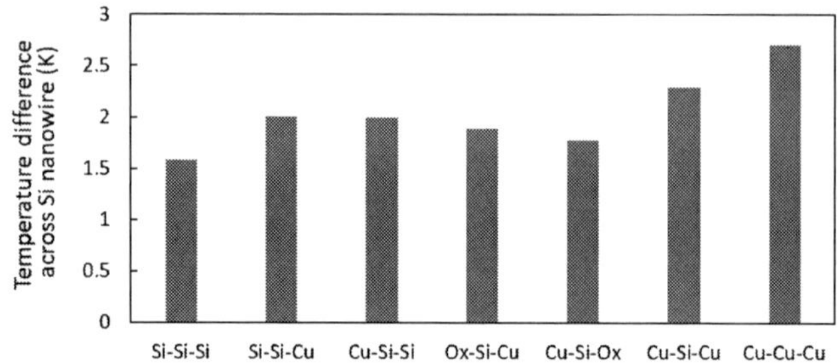

Figure 15. Comparison of temperature difference ΔT across Si-NWs on various types of the substrate (24).

Figure 15 summarizes the temperature differences ΔT across the Si-NW for all samples. Against all expectation, there is no significant difference in ΔT between Ox-Si-Cu and Cu-Si-Ox. The temperature difference in Cu-Si-Cu is larger than those in Ox-Si-Cu and Cu-Si-Ox, indicating that the larger the temperature difference values are established by replacing more Si regions by Cu. The best case is Cu-Cu-Cu, indicating that replacing the whole substrate by Cu is the most effective way to improve the generated power density. The results show that the thermal resistance of the entire substrate should be suppressed as a matter of first priority.

Conclusions

The authors' group has been developing a Si-based micro-thermoelectric generator device which can be fabricated by CMOS technology. Conventional demonstrations of Si-based thermoelectric generator employed long Si-NWs suspended on a cavity to avoid the bypass of the heat current. Contrary, newly proposed device comprises short Si-NWs which are not suspended on a cavity etched on the substrate. It is driven by a steep temperature gradient exuding around a heat flow perpendicular to the substrate, and the power density is scalable by shortening Si-NW to sub-μm length. It has been experimentally confirmed that the thermoelectric power was enhanced by shortening the nanowire, and a very high power generation density corresponding to 12 μW/cm^2 was demonstrated from a temperature difference of 5 K. It is noteworthy that the thermoelectric open circuit voltage was enhanced by shortening the Si-NWs, suggesting that the Seebeck coefficient of the Si-NWs increases as the Si-NW shortens. The physical origin of the Seebeck coefficient enhancement, however, remains an open question. It was found that the device performance largely rests on the suppression of parasitic heat resistance. Further enhancement of the thermoelectric power can be achieved by suppressing the heat resistance of the entire substrate.

Acknowledgments

This work was supported by the CREST projects JPMJCR15Q7 and JPMJCR19Q5, of Japan Science and Technology Corporation (JST), in part by the NIMS Nanofabrication Platform, and in part by AIST-SCR. The authors thank Mr. D. Kawaguchi from Hamamatsu Photonics K.K. for offering stealth dicing, and the members of the CREST project, Prof. Y. Kamakura, Prof. H. Ikeda, Dr. H. Zhang, Dr. S. Hashimoto, Mr. S. Ohba, Mr. S. Asada, and Mr. T. Xu, for their significant contributions to this research.

References

1. H. J. Goldsmid, *Introduction to Thermoelectricity*, Springer 2010.
2. A. I. Boukai, Y. Bunimovich, J. Tahir-Kheli, J. K. Yu, W. A. Goddard III, J. R. Heath, *Nature* **451**, pp. 168-171 (2008).
3. A. I. Hochbaum, R. Chen, R. D. Delgado, W. Liang, E. C. Garnett, M. Najarian, A. Majumdar, P. Yang, *Nature* **451**, pp. 163-167 (2008).
4. G. Pennelli, *Beilstein J. Nanotechnol.* **5**, pp. 1268-1284 (2014).
5. X. Yu, Y. Wang, Y. Liu, T. Li, H. Zhou, X. Gao, F. Feng, T. Roinila, Y. Wang, J. Micromech. Microeng. 22, 105011 (2012).
6. M. Stresser R. Aigner, M. Franosch, G. Wachutka, *Sens. Actuators A, Phys.*, **97-98**, pp. 535-542 (2002).
7. J. Xie, C. Lee, H. Feng, *J. Microelectromechanical Systems* **19**, pp. 317-324 (2010).
8. B. M. Curtin, E. W. Fang, J. Bowers, *J. Electronic Materials* **41**, pp. 887-894 (2012).
9. Y. Li, K. Buddharaju, N. Singh, S. J. Lee, *Journal of Electronic Materials* **41**, pp. 989-992, (2012).
10. A. R. Abramson, W. C. Kim, S. T. Huxtable, H. Yan, Y. Wu, A. Majumdar, C. L. Tien, P. Yang, *J. Microelectromechanical Systems* **13**, pp. 505-513 (2004).
11. K. J. Norris, M. P. Garrett, J. Zhang, E. Coleman, G. S. Tompa, N. P. Kobayashi, *Energy Conversion and Management* **96**, pp. 100–104 (2015).
12. L. Fonseca, J. D. Santos, A. Roncaglia, D. Narducci, C. Calaza, M. Salleras, I. Donmez, A. Tarancon, A. Morata, G. Gadea, L. Belsito, L. Zulian, *Semiconductor Science and Technology* **31**, pp. 084001 (2016).
13. B. Xu, W. Khouri, K. Fobelets, *IEEE Electron Device Letters* **35**, pp. 596-598 (2014).

14. T. Zhang, S.Wu, J. Xu, R. Zhenga, G. Cheng, *Nano Energy* **13**, pp. 433–441 (2015).
15. T. Watanabe, S. Asada, T. Xu, S. Hashimoto, S. Ohba, Y. Himeda, R. Yamato, H. Zhang, M. Tomita, T. Matsukawa, Y. Kamakura, H. Ikeda, *2017 IEEE Electron Devices Technology and Manufacturing Conference (EDTM) Proceedings of Technical Papers*, pp.86-87 (2017).
16. H. Zhang, T. Xu, S. Hashimoto, and T. Watanabe, *IEEE Trans. Electron Devices* **65**, pp. 2016–2023, (2018).
17. T. Zhan R. Yamato, S. Hashimoto, M. Tomita, S. Oba, Y. Himeda, K. Mesaki, H. Takezawa, R. Yokogawa, Y. Xu, T. Matsukawa, A. Ogura, Y. Kamakura, T. Watanabe, *Sci. Technol. Adv. Mater.***19**, pp. 443–453, (2018).
18. M. Tomita, S. Ohba, Y. Himeda, R. Yamato, K. Shima, T. Kumada, M. Xu, H. Takezawa, K. Mesaki, K. Tsuda, S. Hashimoto, T. Zhan, H. Zhang, Y. Kamakuri, Y. Suzuki, H. Inokawa, H. Ikeda, T. Matsukawa, T. Matsuki, T. Watanabe, *Symp. 2018 VLSI Technol. Dig. Tech. Papers*, pp. 93–94 (2018).
19. M. Tomita, S. Ohba, Y. Himeda, R. Yamato, K. Shima, T. Kumada, M. Xu, H. Takezawa, K. Mesaki, K. Tsuda, S. Hashimoto, T. Zhan, H. Zhang, Y. Kamakuri, Y. Suzuki, H. Inokawa, H. Ikeda, T. Matsukawa, T. Matsuki, T. Watanabe, *IEEE Transaction on Electron Devices* **65**, pp. 5180-5188 (2018).
20. T. Xu, S. Hashimoto, S. Asada and T. Watanabe, *14th European Conference on Thermoelectrics (ECT2016),* PC1.11, p.256 (2016).
21. F. Salleh, T. Oda, Y. Suzuki, Y. Kamakura, and H. Ikeda, Appl. Phys. Lett. **105**, 102104 (2014).
22. L. I. Anatychuk, L. P. Blat, "Thermoelectric Phenomena under Large Temperature Gradients," Chapter 3, *Thermoelectric Handbook Micro to Nano*, Ed. by D. M. Rowe, CRC Press (2006).
23. A. N. Grigorenko, P. I. Nikitin, D. A. Jelski, T. F. George, J. App. Phys. **69**, pp. 3375-3377 (1991).
24. K. Shima, M. Tomita, H. Zhang, T. Zhan, T. Matsukawa, T. Matsuki and T. Watanabe, SSDM 2018 Tech. Dig. 429-430 (2018).

ECS Transactions, 89 (3) 111-120 (2019)
10.1149/08903.0111ecst ©The Electrochemical Society

Charge Pumping in Ultrathin SOI Tunnel FETs: Impact of Back-Gate Voltage

C. Diaz Llorente[a], C. G. Theodorou[b], J. -P. Colinge[a], S. Cristoloveanu[b], S. Martinie[a],
C. Le Royer[a], G. Ghibaudo[b] and M. Vinet[a]

[a] CEA-LETI, Université Grenoble Alpes, Grenoble 38000, France
[b] IMEP-LaHC Minatec INP, Université Grenoble Alpes, Grenoble 38000, France

A thorough investigation of trap density (D_{it}) at the interfaces of
FD-SOI Tunnel FET (TFET) devices is presented. Charge
pumping (CP) method confirms a larger average defect
concentration at the top interface of Low-Temperature processed
TFETs than in devices made using a standard high-temperature
process. Back-gate bias is used to avoid probing the back-interface
traps. We demonstrate the link between the CP current
(experiments) and the carrier concentration (TCAD) at the
interfaces during measurement. Simulation results demonstrate that
the unusual variation of CP current in ultrathin FD-SOI depends on
the carriers available for recombination at both front and back
interfaces.

Introduction

The electrical characterization of silicon TFETs fabricated using low-thermal budget
processing has demonstrated an enhancement of the drive current compared to standard
High-Temperature processed (HT) TFETs at high gate voltage ($V_G = -2.0$ V, p-channel
devices) for different drain polarizations. Unfortunately, an important off-current
degradation was observed in the LT TFETs (1). This phenomenon can be explained by
the use of the SPER (Solid-Phase Epitaxial Regrowth technique) (2) to achieve efficient
activation of dopants in the source and drain regions at low-temperature (< 650 °C). One
consequence of using SPER recrystallization at 650°C is the presence of defects at the
interface between the amorphous region and the bottom of the SOI film (3). The presence
of these defects degrades Band-To-Band Tunneling (BTBT) because Trap-Assisted-
Tunneling (TAT) dominates over BTBT at low gate voltages, jeopardizing the possibility
of obtaining a subthreshold swing below 60 mV/dec. The TFETs presented here were
fabricated using either a standard high-temperature (1050 °C) or a low-temperature (600
°C) process. For the first time, we use the charge pumping (CP) method in Silicon On
Insulator (SOI) HT and LT TFETs made in ultrathin silicon (thickness: 11 nm). The EOT
is 1.18 nm and the BOX thickness is 145 nm. TCAD simulations provide a deep
understanding of the back-gate voltage impact on the recombination of carriers. We
explain the special behavior of the charge pumping current and use it for the extraction of
an average density of defects.

Charge Pumping Configuration in FD-SOI TFETs

The charge pumping method (4) consists in applying a periodic signal to the gate terminal in order to switch it from inversion to accumulation and reciprocally. Figure 1 shows the bias configuration used in our Tunnel FET devices for CP measurements. The N^+ region is grounded (or reverse biased) and a "square" pulse is applied to the front gate terminal (5,6). The charge pumping current (I_{CP}) is measured at the P^+ contact when the gate is pulsed at frequency f and is given by (7):

$$_{CP} = q^2 f A \overline{D}_{it} \Delta\phi_S \tag{1}$$

where A is the effective channel area of the transistor, $\overline{D}_{it}$ is the average density of defects and $\Delta\Phi_S$ is the surface potential range scanned within the bandgap. Equation [1] establishes that I_{CP} is proportional to both D_{it} and frequency. Therefore, it is possible to obtain the average density of defects from the slope of the $I_{CP}(f)$ curve (8).

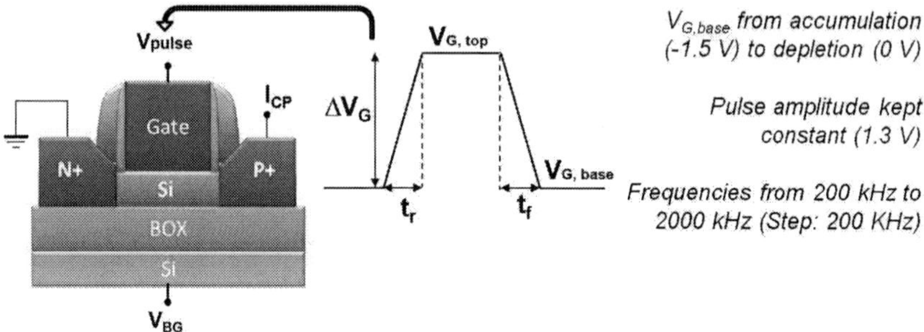

Figure 1.   Charge pumping method configuration in TFET structures and setup characterization with "square" pulses applied to the front-gate terminal to generate the charge pumping current.

The charge pumping method requires the presence of majority and minority carriers since I_{CP} current results from the recombination of detrapped electrons in the channel region with holes during the cycle of a pulse. In SOI layers the channel region is separated from the substrate by the BOX. Moreover, the body region is thin and lightly doped ($\sim 10^{15}$ cm^{-3}), such that the amount of available majority carriers is limited, which affects the net charge pumping current that is measured. Fortunately, in our FD-SOI TFETs both types of carriers are available because the doping polarities in the source and drain regions are opposite (9). These terminals serve as reservoirs that can instantly supply any required amount of electrons and holes.

Experimental results of charge pumping current versus pulse base level ($V_{G,base}$) show a well-defined bell shape characteristics (Figure 2). This is explained by the position of the pulse with respect to the flat band voltage (V_{FB}) and threshold voltage (V_{TH}). In region 3, the pulse base (lowest) level forms an accumulation channel ($V_{G,base} < V_{FB}$) and the top level induces strong inversion ($V_{G,top} > V_{TH}$). In this state then entire bandgap is scanned and all the defects will be filled with electrons (holes) at the top (bottom) of the pulse. This bias condition generates the maximum charge pumping current.

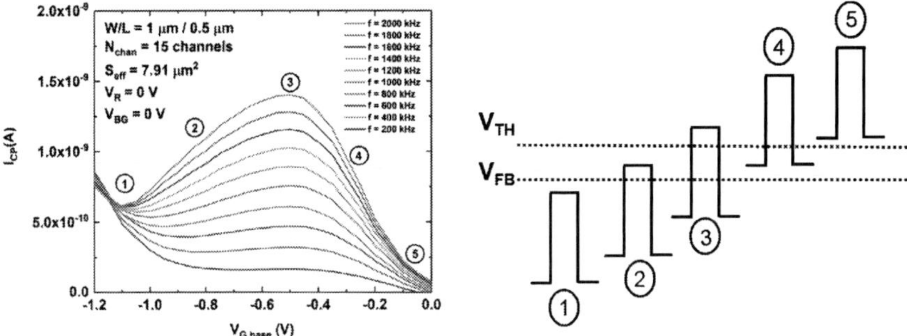

Figure 2. $I_{CP}(V_{G,base})$ curves showing the recombination current measured at given pulse amplitude and for different frequencies, with a schematic of the gate pulse at different phases: the maximum charge pumping current is obtained in phase 3.

Impact of Back-Gate Voltage on Charge Pumping Current

In FD-SOI structures, the top and bottom interfaces are extremely close. This means that the pulse applied on the front-gate can sweep the whole channel film from accumulation to inversion. Therefore, carriers from the top region could also be trapped in defects located at the back interface, enabling a net contribution to the I_{CP} current and thus, providing a non-accurate value of D_{it}. This clearly indicates that it is impossible to be certain that the measured I_{CP} current is due to the presence of defects at the gate oxide/channel interface when no back-gate voltage is applied (Figure 3). For this reason it is mandatory to apply a back-gate voltage (V_{BG}) to obtain accurate front-channel measurements. This way it is possible to avoid measuring any I_{CP} contribution from the back interface.

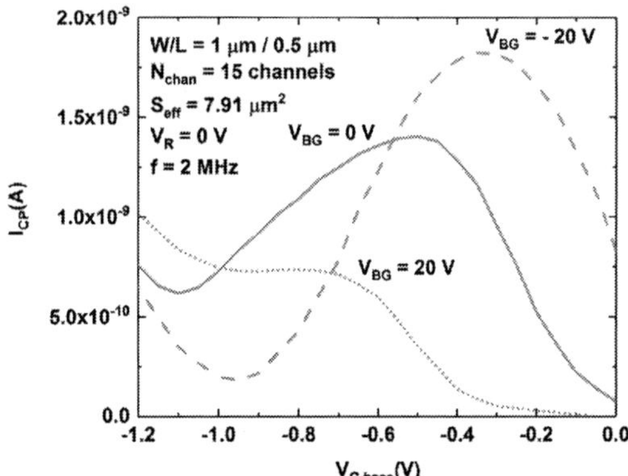

Figure 3. Charge pumping current evolution versus pulse base level at different back-gate voltages in a p-type TFET device (T_{Si} = 11 nm, T_{BOX} = 145 nm).

For V_{BG} = - 20 V an accumulation layer of holes is formed at the bottom interface. This does not allow the front-gate pulse to induce trapping of electrons at the back interface. As a result the contribution of the bottom traps to I_{CP} is reduced. However, the electrostatically doped P-type layer created at the bottom of the SOI channel decreases the channel potential. Therefore, it is necessary to apply a higher gate voltage to the front gate to generate an inversion layer and, as a result, threshold voltage is increased. This is shown in Figure 3, where a shift of the I_{CP} curve towards less negative values of the pulse base level for V_{BG} = - 20 V is observed.

The influence of the substrate polarization on the CP current can be found in the literature. In thick SOI films (450 nm), Wouters *et al.* (10) reported that there is no significant change in the $I_{CP}(V_{G,base})$ curves, for different values of applied back-gate voltage. However, Ouisse *et al.* (8) demonstrated that for SOI gated p-i-n diodes with a silicon film thickness below 300 nm there was a clear variation of the maximum charge pumping current when applying a back-gate polarization. In particular, in depletion mode (V_{BG} = 0 V) an increase of the $I_{CP,max}$ was observed, unlike the case of the 450 nm thick film (10). A contribution of the back interface to the charge pumping current was the consequence of scanning the bandgap of the back interface. This behavior was also confirmed in a later investigation done on 100 nm thick FDSOI MOSFETs (11).

In our TFETs with a silicon film thickness of 11 nm, there is a clear dependence of $I_{CP,max}$ on back bias. Specifically, there is an increase of the charge pumping current at V_{BG} = - 20 V, when the bottom interface is theoretically deactivated. However, when a zero back-gate voltage is applied, the front-gate pulse should cause a sweep of the back surface potential between accumulation and inversion and thus, add a contribution of defects at the bottom of the channel to the CP current arising from front-interface defects. A reasonable explanation is that when applying a V_{BG} < 0, an extra depletion region is generated in the middle and bottom channel regions coming from the back gate, since the silicon film is very thin (Figure 4). Therefore, when pulsing the front gate the defects present in this region and not only at the top interface, will also participate in the recombination process.

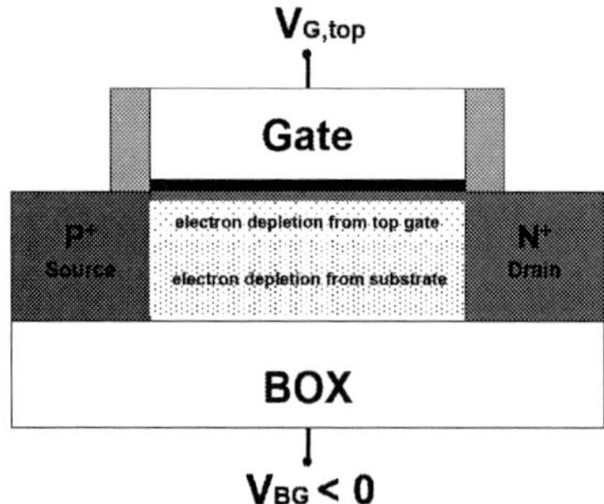

Figure 4. Cross-section of the schematic TFET structure showing the depletion region generated by the pulse top level and the depletion due to the substrate polarization. Defects in this region can trap electrons from the top inversion layer, increasing the charge pumping current.

Density of Interface States in LT and HT TFETs

In this section we present the average defect density obtained via the CP method at the front interface, when applying a back-gate voltage of − 20 V for both HT and LT Tunnel FET devices.

TABLE I. Average defect density obtained at the front-interface for high-temperature (HT) and low-temperature (LT) processes. $L_G = 0.5\,\mu m$, $V_{G,base} = -0.65$ V and $\Delta V_G = 1.3$ V.

			HT (1050ºC)	LT (650ºC)
Width (µm)	Nº channels	A_{eff} (µm^2)	N_{it} (cm^{-2}eV^{-1})	N_{it} (cm^{-2}eV^{-1})
1.0	15	7.91	$7.0\cdot10^{10}$	$1.5\cdot10^{11}$
0.5	30	8.33	$7.8\cdot10^{10}$	$8.8\cdot10^{11}$
0.2	50	6.38	$4.8\cdot10^{10}$	$7.2\cdot10^{10}$
0.1	75	5.81	$6.4\cdot10^{10}$	$1.1\cdot10^{11}$
0.05	75	3.94	$4.1\cdot10^{10}$	$4.2\cdot10^{11}$

From Table I we observe a higher defect density in High-Temperature TFETs than in their low-temperature counterparts. This is a first confirmation to attribute the current enhancement of devices characterized in (1) (with $L_G = 500$ nm and $W = 1.0$ µm) to TAT due to a higher presence of traps located in the semiconductor-insulator interfaces or in the bulk of the semiconductor, resulting in a generation current that degrades the subthreshold slope. In addition, a second confirmation was made based on low-frequency noise analyses (12). These noise results confirm the presence of a higher defect concentration of traps located at the source junction in the low-temperature process Tunnel FETs.

Charge Pumping: Density of Carriers and I_{CP} Current

In order to understand the physics of the CP measurements, we have performed 2D TCAD study based on the simulation of the carrier density at the top and bottom interfaces for different back-gate voltages. The aim is to confirm there is some correlation between the measured I_{CP} and the carrier concentrations.

The hypothesis we try to verify is that that I_{CP} current is limited by the concentration of carriers that are able to recombine. So, if the difference of electron and hole concentrations available in the film during a charge pumping cycle is significant, the recombination will be proportional to the lowest density of carriers available (either holes of electrons). This suggests the necessity of a balance with respect to hole and electron concentrations to yield reliable charge pumping measurements. For the TFET simulation deck we have used the same geometry and parameters as in (1). However, the source and drain junctions are located at the gate edge. The pulse applied in the front-gate is simulated by two different voltage conditions: the pulse base level is − 0.65 V and the top level is + 0.65 V.

$V_G = -0.65$ V

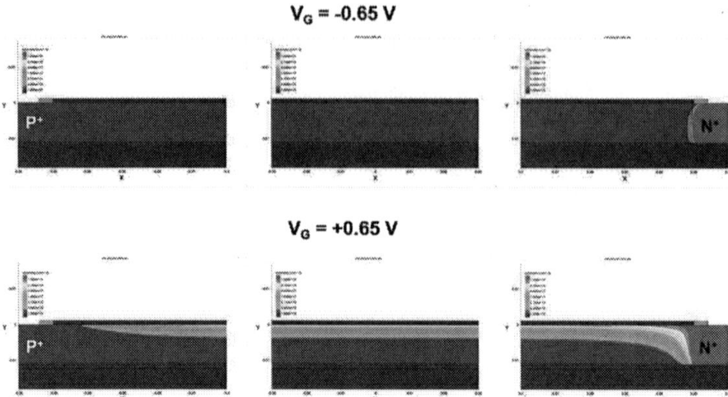

Figure 5. Electron density in different regions of the TFET structure (anode, channel and cathode): (top) at pulse base level the whole film thickness is depleted of electrons; (bottom) at pulse top level the electrons are located at the top-interface ($V_{BG} = -20$ V).

More specifically, for $V_{BG} = -20$ V and $V_{G,top} = +0.65$ V the electron density is located at the top surface creating an inversion layer (Figure 5), while the back interface is depleted of electrons (concentration lower than 10^{15} cm^{-3}). For $V_{G,base} = -0.65$ V, two layers of holes are formed at the channel interfaces as shown in Figure 6. Considering both mechanisms at the same time, during the falling edge of the pulse, the concentration of electrons that could be trapped by defects at the bottom interface is at most equal to 10^{15} cm^{-3}. Although the concentration of holes is significantly higher, the recombination at the back interface is negligible because it is proportional to the lowest concentration of either carrier (here electrons). So, traps at the back-interface are rather empty of electrons and their contribution to the I_{CP} current is low.

$V_G = -0.65$ V

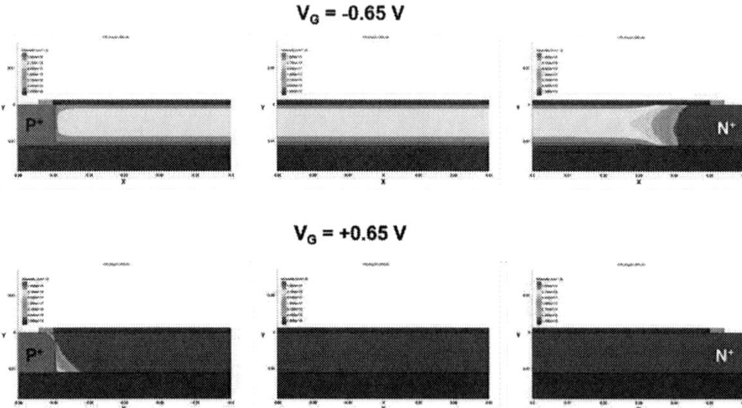

Figure 6. Hole density in different regions of the TFET structure (anode, channel and cathode): (top) at pulse base level, front and back interfaces are filled with holes; (bottom) at pulse top level the channel is depleted of holes ($V_{BG} = -20$ V).

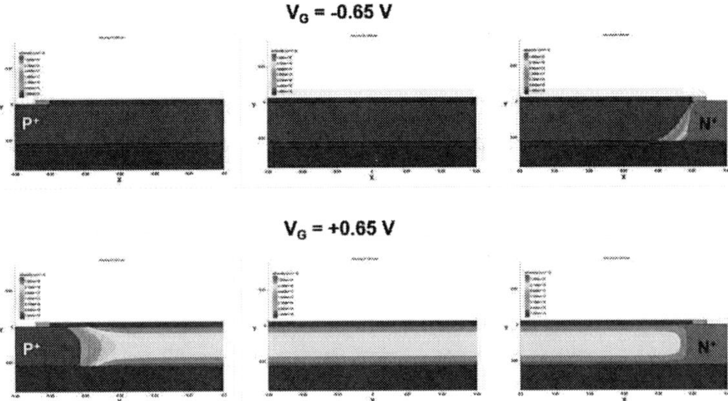

Figure 7. Electron density in different regions of the TFET structure: (top) at pulse base level the whole film thick is depleted of electrons; (bottom) at pulse top level the electrons are located at both top and back interfaces (V_{BG} = + 20 V).

The situation is different for V_{BG} = + 20 V. In particular at $V_{G,top}$ = + 0.65 V (Figure 7), electrons are present at each interface. However, when switching the gate voltage to $V_{G,base}$ = - 0.65 V, it is possible to observe complete depletion of electrons in the whole channel region. At the pulse base level, holes are found only at the top surface, while the back interface is depleted of holes, deactivating the possibility of recombination (Figure 8). Therefore, the contribution to the I_{CP} from the back interface is negligible, due to the impossibility of scanning the entire band gap. Finally, when V_{BG} = 0 V, electron and hole concentrations at the back interface are similar in value and participate in the recombination process.

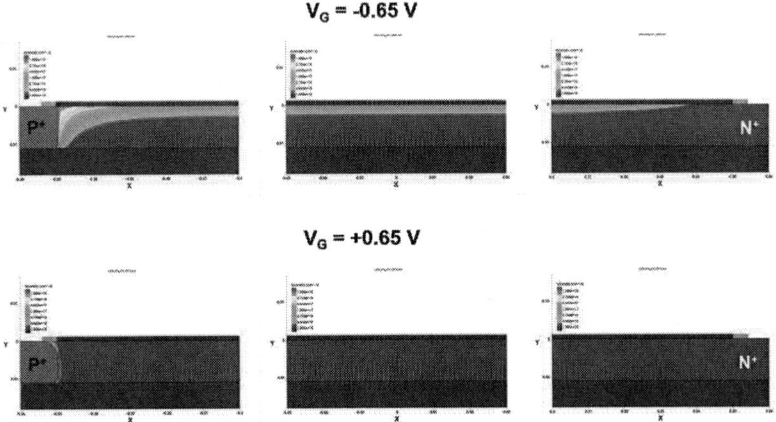

Figure 8. Hole density in different regions of the TFET structure: (top) at pulse base level only the front interface is filled with holes; (bottom) at pulse top level the channel is depleted of holes (V_{BG} = + 20 V).

From the TCAD simulations we assume that charge pumping current is proportional to the minimum carrier concentration (min[electron concentration, hole concentration]). In addition, carrier concentrations lower than 10^{15} cm^{-3} in the top/bottom channel regions are considered too low to provide a significant contribution to the I_{CP} current.

V_{BG} (V)	V_G (V)	Top h (cm^{-3})	Top e (cm^{-3})	Bottom h (cm^{-3})	Bottom e (cm^{-3})	Rec. conc. at Top (cm^{-3})	Rec. conc. at Bottom (cm^{-3})	Sum of e contrib. (cm^{-3})	Sum of h contrib. (cm^{-3})
-20	-0.65	$4.6 \cdot 10^{18}$	0	$1.3 \cdot 10^{19}$	0	$6.2 \cdot 10^{17}$	$1.4 \cdot 10^{10}$		$1.8 \cdot 10^{19}$
	0.65	0	$6.2 \cdot 10^{17}$	$1.1 \cdot 10^{9}$	$1.4 \cdot 10^{10}$			$6.2 \cdot 10^{17}$	
0	-0.65	$4.0 \cdot 10^{18}$	0	$3.6 \cdot 10^{17}$	0	$4.0 \cdot 10^{18}$	$3.6 \cdot 10^{17}$		$4.4 \cdot 10^{18}$
	0.65	0	$7.5 \cdot 10^{18}$	0	$4.3 \cdot 10^{17}$			$7.9 \cdot 10^{18}$	
20	-0.65	$7.4 \cdot 10^{16}$	0	$1.6 \cdot 10^{9}$	$8.6 \cdot 10^{9}$	$7.4 \cdot 10^{16}$	$1.6 \cdot 10^{9}$		$7.4 \cdot 10^{16}$
	0.65	0	$8.2 \cdot 10^{18}$	0	$1.3 \cdot 10^{19}$			$2.1 \cdot 10^{18}$	

Figure 9. Concentration of carriers obtained at the top and back interfaces. Two cases are studied: (i) only the top contribution is considered and (ii) both top and bottom contributions are considered.

Simulation results indicate that the variation of the charge pumping current depends of the carriers available for recombination at the front and back interfaces. Therefore either one of these concentrations decreases when the back-gate voltage is more negative (fewer electrons) or more positive (fewer holes), not only at the back but also at the top interface. In Figure 9, the recombination concentrations at the top and at the bottom respectively (blue columns), are the result of the recombined electron/hole pairs from the top and bottom isolated from each other. Analyzing the bottom region separately from the top, we observe that when applying a back-gate bias, the total amount of recombining carriers is much lower than 10^{15} cm^{-3} (red arrows). This indicates that the back interface is deactivated. However, at V_{BG} = 0 V, electrons and holes are in the same order of magnitude ($\sim 10^{17}$ cm^{-3}, green arrow), suggesting that the back interface will contribute to the I_{CP} current.

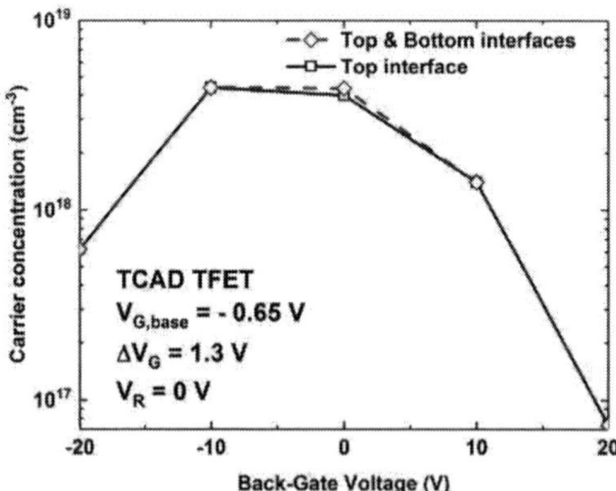

Figure 10. TCAD simulation of minimum carrier concentration for variable back-gate voltages at $V_{G,base}$ = - 0.65 V. No remarkable difference when only the top concentration is considered, with respect to the top and bottom concentration contributions (12).

Figure 10 shows the minority carrier concentration (which governs the recombination rate and I_{CP}) for different applied back-gate voltages. The curve of the carrier concentration presents the same shape as the one obtained for the experimental charge pumping current (12), explaining the recombination mechanism. Same consistent results between experiments and simulations have been obtained for different $V_{G,base}$ voltages respectively (- 0.35 V and -0.85 V).

Conclusions

Our updated charge pumping method is applied to determine and compare the density of traps in high-temperature and low-temperature FDSOI Tunnel FETs. The application of a back-gate voltage allows one to obtain a more accurate calculation of D_{it}, because the pulse applied on the front-gate terminal cannot sweep the whole film thickness from accumulation to inversion and vice-versa. Experimental results exhibit a higher defect density in LT TFETs ($\sim 10^{11}$ cm^{-2}eV^{-1}) than the HT devices ($\sim 10^{10}$ cm^{-2}eV^{-1}). This evidence suggests that TAT dominates over BTBT lateral tunneling at low gate voltages, deactivating the possibility to obtain a steep subthreshold slope. To confirm this analysis, we have performed TCAD simulations based on the evolution of the carrier concentration at different pulse base levels, while keeping the amplitude of the pulse constant. For $V_{G,base} = - 0.65$ V, simulations of carrier concentration show the same trend than the experimental charge pumping current. This clearly indicates the importance of the balance between electron and hole concentrations. If there is a difference higher than one order of magnitude between these two concentrations, the recombination process in the CP method (and therefore the net I_{CP} current) will be proportional to the lowest concentration of either carrier. Therefore, we will obtain a non-accurate value of the average defect density.

Acknowledgments

This work is partly funded by the French Public Authorities through NANO 2017 program and EQUIPEX FDSOI11.

References

1. Diaz Llorente C., Le Royer C., Batude P., Fenouillet-Beranger C., Martinie S, Lu C.-M.V., et al., New insights on SOI Tunnel FETs with low-temperature process flow for CoolCube^{TM} integration. Solid-State Electronics. 2018; 144:78-85.
2. Fenouillet-Beranger C., Batude P., Brunet L., Mazzocchi V., Lu C.-M.V., Deprat F., et al., Recent advances in low temperature process in view of 3D VLSI integration. IEEE SOI-3D-Subthreshold Microelectronics Technology Unified Conference (S3S). 2016. p. 1-3.

3. Sklenard B., Xu C., Batude P., Previtali B., Tabone C., Rafhay Q., et al., FDSOI devices: A solution to achieve low junction leakage with low temperature processes ($\leq$ 650°C). International Conference on Ultimate Integration on Silicon (ULIS). 2012. p. 169-72.

4. Brugler J.S., Jespers P.G.A., Charge pumping in MOS devices. IEEE Transactions on Electron Devices. 1969; 16(3):297-302.

5. Groeseneken G., Maes H.E., Beltran N., Keersmaecker R.F.D., A reliable approach to charge-pumping measurements in MOS transistors. IEEE Transactions on Electron Devices. 1984; 31(1):42-53.

6. Li Y., Wang G., Ma T.P., A front gate charge pumping technique for measuring both interfaces in fully depleted SOI/MOSFETs. Proceedings of International Electron Devices Meeting. 1995. p. 643-6.

7. Seghir K., Cristoloveanu S., Jerisian R., Oualid J., Auberton-Herve A-J., Correlation of the leakage current and charge pumping in silicon on insulator gate-controlled diodes. IEEE Transactions on Electron Devices. 1993; 40(6):1104-11.

8. Ouisse T., Cristoloveanu S., Elewa T., Haddara H., Borel G., Ioannou D.E,. Adaptation of the charge pumping technique to gated p-i-n diodes fabricated on silicon on insulator. IEEE Transactions on Electron Devices. 1991; 38(6):1432-44.

9. Le Royer C., Villalon A., Martinie S., Nguyen P., Barraud S., Glowacki F., et al., Experimental investigations of SiGe channels for enhancing the SGOI tunnel FETs performance. 2015 Joint International EUROSOI Workshop and International Conference on Ultimate Integration on Silicon. 2015. p. 69-72.

10. Wouters D.J., Tack M.R., Groeseneken G.V., Maes H.E., Claeys C.L., Characterization of front and back Si-SiO$_2$ interfaces in thick- and thin-film silicon-on-insulator MOS structures by the charge-pumping technique. IEEE Transactions on Electron Devices. 1989; 36(9):1746-50.

11. Li Y., Ma T.P., . A front-gate charge-pumping method for probing both interfaces in SOI devices. IEEE Transactions on Electron Devices. 1998; 45(6):1329-35.

12. Diaz-Llorente C., Colinge J.-P., Le Royer C., Vinet M., Theodorou C.G., Cristoloveanu S., et al., Impact of Low-Temperature Coolcube[TM] Process on the Performance of FDSOI Tunnel FETs. IEEE SOI-3D-Subthreshold Microelectronics Technology Unified Conference (S3S). 2018. p. 1-3.

ECS Transactions, 89 (3) 121-131 (2019)
10.1149/08903.0121ecst ©The Electrochemical Society

Roles of Inner and Outer Fringe and Asymmetric Coupling Effect in Concentric Double-MIS(p) Tunneling Diodes

Yu-Hsuan Chen[a] and Jenn-Gwo Hwu[b]

[a] Graduate Institute of Electronics Engineering, National Taiwan University, Taipei, Taiwan
[b] Graduate Institute of Electronics Engineering / Department of Electrical Engineering, National Taiwan University, Taipei, Taiwan
E-mail: jghwu@ntu.edu.tw

During the studies of coupling effects between the outer ring and inner circle in a concentric structure of double-metal-insulator-semiconductor tunnel diode, we have proposed an asymmetric coupling effect to account for the distinctive difference between two opposite operations. However, further investigation is needed to make our mechanism more robust. In this work, we designed two sets of ring-shaped and concentric MIS tunnel diode lithographic patterns with a fixed inner fringe in the first set and a fixed outer fringe in the second set. From both the experimental and TCAD-simulated results, we have found that the major role in grabbing charges is the outer fringe of a ring-shaped MIS structure and that this outer fringe-dominated characteristic accompanied with the asymmetric coupling effect results in the observable difference in two contrary operation modes where the inner circle and outer ring take turns to play as the control gate.

Introduction

It has been known that the saturated current a of a Metal-Insulator-Semiconductor(p) tunnel diode (TD) is periphery-dependent because of the magnified fringing-field effect. This trait has been found and discussed in (1) that if we divide the saturation currents of several single MIS(p) TDs with respect to their perimeters, the currents will be merged accordingly. Furthermore, another renowned characteristic of MIS(p) is that its saturation current is strongly dependent on the Schottky barrier height. The hole current can be expressed as

$$I_h = A^* A_{eff} P_t T^2 \exp(\frac{-q\varphi_{BP}}{kT}) \exp(\frac{q\Delta\varphi_{BP}}{kT}) \qquad [1]$$

(2), where I_h is the hole current, A^* is the effective Richardson constant, A_{eff} is the effective area for the hole flux, P_t is the tunneling probability, T is the temperature of the device, q is the electron charge, k is the Boltzmann's constant, φ_{BP} is the Schottky barrier height without excess minority supply and $\Delta\varphi_{BP}$ is the change of Schottky barrier height due to excess minority supply. Due to its high sensitivity to the change of minority carriers from the surroundings, MIS(p) shows noticeable results towards light or bias-added neighboring gate in a concentric structure. In previous research of double MIS(p) TDs (3), it was shown that through the coupling effect from the neighboring gate with adequate bias, the dark current of central MIS(p) TD can be lowered and the light current can be improved and therefore an excellent light-to-dark ratio can be achieved.

A detailed discussion on the amount of coupling contribution from the inner and outer fringe of the outer ring-shaped MIS TD is still lacking. In this work, two sets of ring-shaped and concentric MIS TD patterns are designed. The first set is made with the outer fringe R_{Oo} fixed and inner fringe R_{Oi} varying, and the second set is designed contrarily.

In this work, we found that the dominant factor in grabbing charges in a double-fringed ring structure is the outer fringe. This discovery accompanied with the asymmetric coupling effect helps us to describe the difference in the two operations in our previous research (4) that discusses the contrary operation when we use the inner circle to play as the control gate and the outer ring to serve as sensor and achieve higher sensitivity while under lower bias.

Experimental Details

A boron-doped p-type silicon wafer with the resistivity of 1-10 Ω.cm was used as the substrate. The wafer was first cleaned by a standard Radio Corporation of America (RCA) clean process. Then the whole wafer was oxidized by anodic oxidation in deionized water with platinum placed as the inert electrode on the cathode side in order to obtain an ultra-thin oxide on silicon. The oxidation process was conducted under a dc voltage of 15 V for 8 min at room temperature. The wafer was tilted with respect to the flat platinum electrode to obtain various oxide thicknesses on one single wafer. The main factor behind these varying thicknesses is the magnitude of the electric field corresponding to different distances between the tilted wafer and the platinum plate. The oxide and the interface quality were then improved by rapid thermal annealing (RTA) at 950 $^\circ$C for 15 seconds. An aluminum film of about 200 nm was deposited by thermal evaporation as the top contact. Photolithography and wet etching followed to define the concentric circle pattern. Finally, the back native oxide was removed by buffer oxide etchant (BOE) and the 200 nm aluminum back contact was also deposited by evaporation. The current characteristics were measured by an Agilent B1500A system. All the measurements of electrical characteristics are carried out at room temperature, which is about 300 K.

Results and Discussion

(I)   Experimental part

The schematics of the top view, cross-section and parameter table of ring-shaped and concentric double-MIS TD are shown in Fig. 1 (a) and (b), respectively. The radius of inner and outer fringe of the outer ring are named as R_{Oi} and R_{Oo}. Besides, "V_{ring}" denotes the bias of the ring-shaped MIS TD in Fig. 1 (a) and the notations "V_I" and "V_O" represent the bias of inner and outer electrodes in the concentric MIS TD in Fig. 1 (b). Both parameter tables for the ring-shaped and concentric double-MIS TD are shown with a noticeable design that are divided into two sets. Set 1 consists of three samples with the same radius of the outer fringe of the outer ring (R_{Oo} = 600 μm) while the three samples in set 2 have their inner fringe of the outer ring (R_{Oi} = 115 μm) fixed. The gap between the outer ring and the inner circle of the concentric double-MIS TD in Fig. 1 (b) are all equal to 30 μm and therefore the radius of the inner circle would be 85 μm.

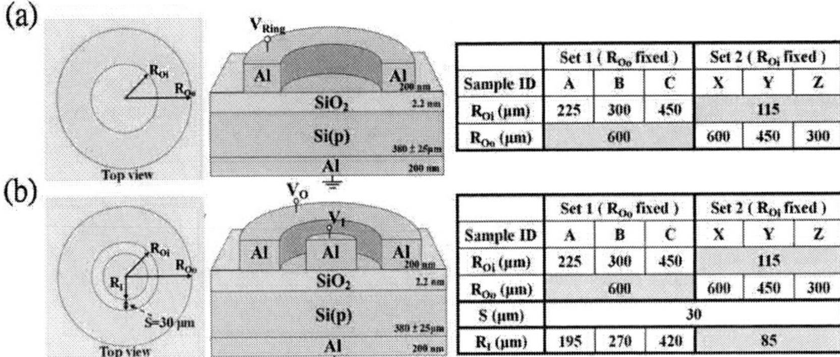

	Set 1 (R_{Oo} fixed)			Set 2 (R_{Oi} fixed)		
Sample ID	A	B	C	X	Y	Z
R_{Oi} (μm)	225	300	450	115		
R_{Oo} (μm)	600			600	450	300

	Set 1 (R_{Oo} fixed)			Set 2 (R_{Oi} fixed)		
Sample ID	A	B	C	X	Y	Z
R_{Oi} (μm)	225	300	450	115		
R_{Oo} (μm)	600			600	450	300
S (μm)	30					
R_I (μm)	195	270	420	85		

Figure 1. Schematic diagrams of the top view, cross section and the parameter table of (a) ring-shaped and (b) concentric MIS TD structure. Here we use R_{Oi} and R_{Oo} to represent the radius of the inner and outer fringe of outer electrode in Fig. 1 (a), respectively. Within the concentric MIS TD structure in Fig. 1 (b), the inner circle is named as inner electrode (I) and the outer ring is outer electrode (O).

Figure 2 shows the result of the I-V curves of the ring-shaped MIS TD in Fig. 1 (a). It is noted that the distinct difference lies in the saturated light current part where the result of set 1 (Fig. 2 (a)) merges perfectly while that of set 2 (Fig. 2 (b)) do not. We further divide the currents in Fig. 2 (a) with their area ($J_{A,ring}$) and obtain merged currents in the accumulation region in Fig. 3 (a), indicating they are uniform TDs with same oxide thickness. Then, we divide Fig. 2 (a) with their respective total perimeter ($J_{P,ring}$) and find that the saturated dark currents merged better now in Fig. 3 (b). This can be attributed to the similar perimeter-dependent phenomenon as in (1) where the saturated currents of several single MIS TDs with different sizes are merged after they are divided by their perimeters, respectively. That result is attributed to the magnification of the fringing field accompanied by the fact that the current of the MIS(p) is related to the amount of Schottky barrier modulation between oxide and silicon. Combining the results above, it is inferred that both the inner and outer fringes contribute to the saturated current of a ring-shaped MIS TD at dark, however their respective amounts are by intuition unequal owing to their different symmetry of electric field. Moreover, with the addition of external light the dominant term in the saturated current becomes that amount of charges grabbed by the outer fringe through a diverging field so that the light current shows great consistency for devices with the same R_{Oo} (Fig. 2 (a)). In this work, white light-emit-diodes with 150 lux are used as the light source to represent common environmental light in our everyday life.

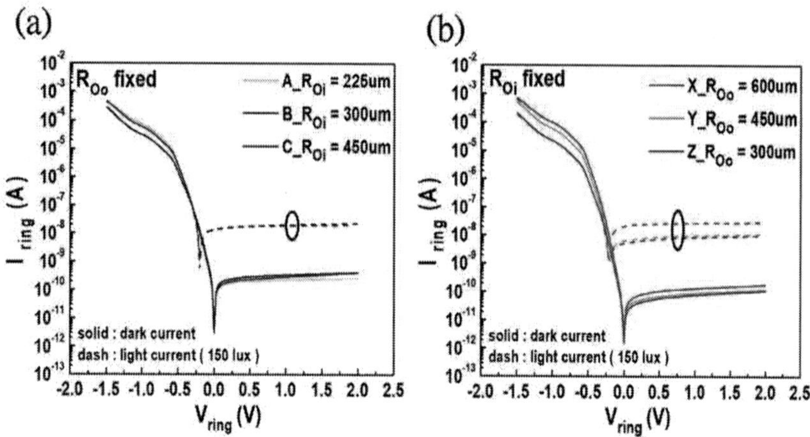

Figure 2. I-V curves under dark and illuminated environments of ring-shaped MIS TD in Fig. 1 (a). Fig (a) and (b) are corresponding to the results of set 1 and set 2, respectively.

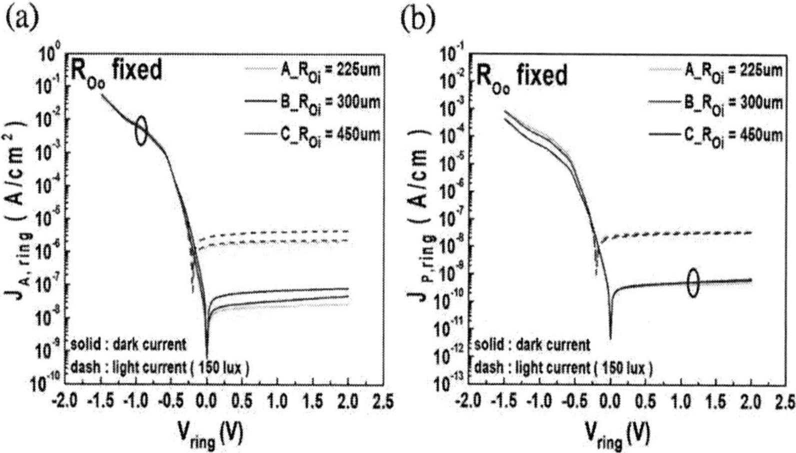

Figure 3. (a) J_A and (b) J_P of ring-shaped MIS TD currents of set 1 in Fig. 1 (a) where R_{Oo} is fixed and R_{Oi} varying.

In the following analysis, we examine the concentric double-MIS TD structure with a 30 μm gap under two operation modes. One is the case where the inner circle functions as the gate and the outer ring plays as the sensor [i.e., inner gate outer sensor (IGOS)], as shown in Fig. 4 (a). In the opposite case (Fig. 4 (b)), the inner circle serves as sensor while the outer ring works as the control gate [i.e., inner sensor outer gate (ISOG)]. It was found in our previous work (4) that the extent of the coupling effect between the inner device and the outer ring is unequal and is proposed as an asymmetric coupling effect (ACE). This difference is reflected and can be observed in the different rightward-shifting extent of the sensor current as the control gate bias increase from 0 to 1.0 V with an increment of 0.2 V in these two operation modes. The schematic concepts of the ACE are shown in the inset of Fig. 4. By utilizing this asymmetric coupling effect, a higher sensitivity can be achieved at a lower voltage bias, and therefore become more power-efficient in IGOS operation. Fig. 4 (a) and (b) show the I-V curves of sample A in IGOS and ISOG operations. We observed that in the IGOS operation, the light currents under different V_{IG} are nearly fixed. This can be interpreted as a combined result of the asymmetric coupling effect and the above-mentioned inference that the main role in grabbing carriers under illumination is for the outer fringe of the ring. Also, similar results are found among samples with 6 different outer ring-to-inner circle area ratios ranging from 48 to 4. Contrarily, in the ISOG operation, both of the extent of dark and light current influenced by outer control bias are much larger owing to the strongly besieging field.

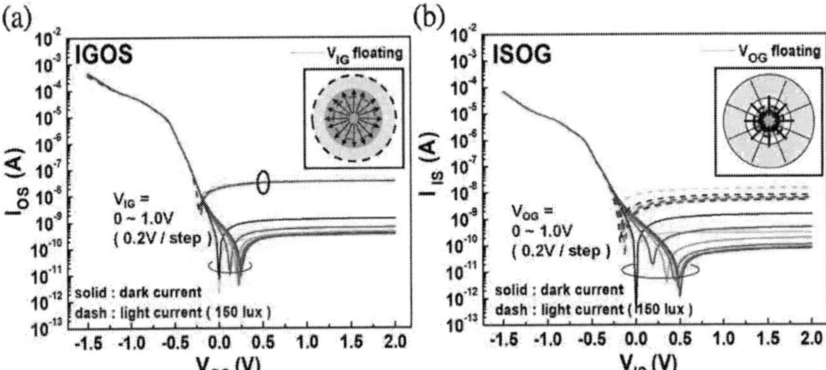

Figure 4. I–V curves of sample A in Fig. 1 (b) under operation when (a) the inner circle serves as a control gate and the outer ring is a sensor (i.e., IGOS) and (b) the inner circle plays as a sensor role and the outer ring is the control gate (i.e., ISOG). The notation "V_{IG}" means the voltage of inner control gate and "V_{OS}" means the voltage of outer sensor in (a), and vice versa in (b).

Finally, Fig. 5 shows the I-V curves of the sensor in a concentric MIS TD structure with R_{Oo} fixed and the total power consumption of (a) IGOS and (b) ISOG operations under a control bias of 1.0 V. It is noted that as the outer ring-to-inner circle area ratio increases (from 0.9 for sample C to 8 for sample A), the light-to-dark current ratio rises while the total power consumption becomes lower and therefore more efficient in both IGOS and ISOG operation modes. Further studies are on the progress to discuss the moderate outer ring-to-inner circle area ratio that can optimize the overall performance.

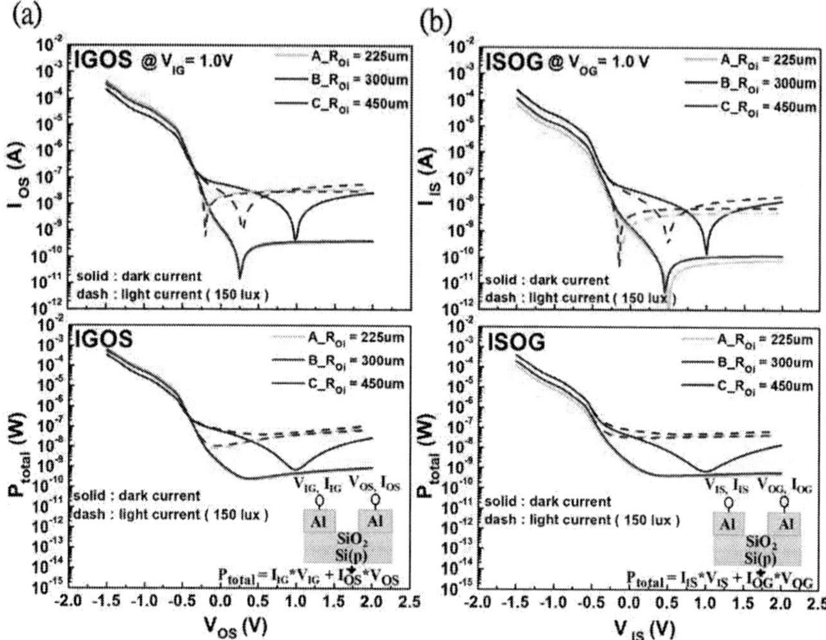

Figure 5. I-V curves of sensors under control bias = 1.0V and total power consumption for set 1 of concentric MIS TD structure in (a) IGOS and (b) ISOG operation mode.

(II)   Simulated part

In the following discussion we use SILVACO TCAD to simulate the concentric double-MIS(p) structure and study the behavior of the electric field and the electron current density under both ISOG and IGOS operation modes. Figure 6 shows the structure of the concentric double-MIS(p) TD. The purple segment on the top of the whole structure is the aluminum electrode with a 250 nm thickness. Below the top electrode and colored as blue is the 2 nm SiO_2. Finally, the yellow part is p-type silicon with a bottom electrode as ohmic contact. The following simulations are all under dark environment.

Figure 6. TCAD-simulated structure of the concentric double-MIS(p) TD. The thickness of the top aluminum electrode is set as 250 nm and the oxide thickness is equal to 2 nm.

Figure 7 shows the electric field distribution extracted from the location with a depth of 1.5 nm below the SiO_2/Si interface for both of the (a) IGOS and (b) ISOG operation modes. The bias of the control gate is set as 1.5 V and the sensor remains floating. This figure is aimed at discussing the unequal coupling extent involved when the outer ring and inner circle plays as the control gate. From Fig. 7 (a) and (b), we can see that the red color, which represents the strongest electric field, happens on the control gate part (in Fig. 7 (a) this will be inner circle and (b) will be outer ring). However, it is noted that an observable difference lies in the sensor part in these two operation modes. The color of the sensor part in Fig. 7 (b) (the inner circle) is much darker than that in Fig. 7 (a) (the outer ring), which means that although under same control gate bias, the coupling degree exerted from the outer ring on the inner circle will be much stronger. This result can be linked to the asymmetric coupling effect as we mentioned above.

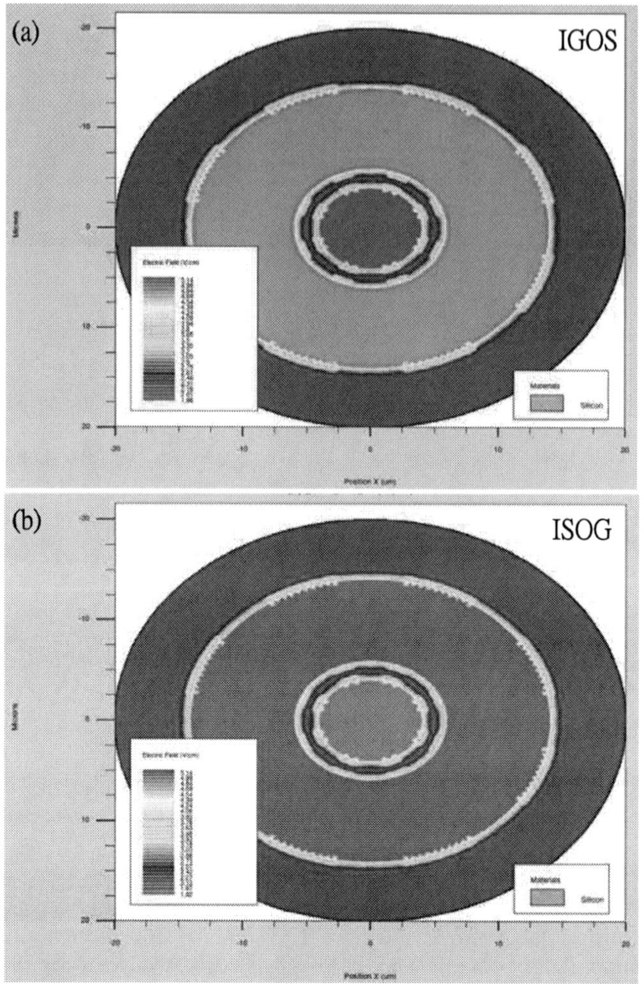

Figure 7. Top view of the electric field distribution in the location with a depth of 1.5 nm below the $SiO_2/Si(p)$ interface under (a) IGOS and (b) ISOG operation modes. The bias of the control gate is equal to 1.5 V and the sensor is floating.

Figure 8 shows the electron current density extracted at the same depth as in Fig. 7. It is noted that most of the current flow along the fringe in both of Fig. 7 (a) and (b), which meets the well-known trait of the single MIS(p) and explains why the saturated current of the single MIS(p) TD with different sizes can be merged after being divided by their fringes. Furthermore, from Fig. 8 (b) we can see that the current density along the outer fringe of the outer ring seems to be higher. In order to further investigate it, we extract the number along the chosen cutline.

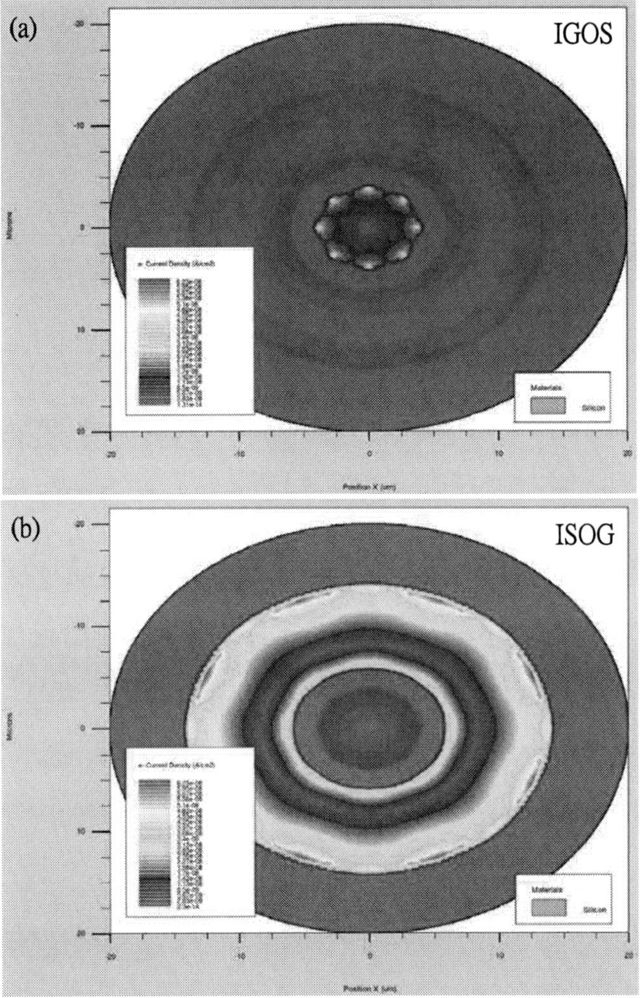

Figure 8. Top view of the electron current density under (a) IGOS and (b) ISOG operation modes. The bias of the control gate is equal to 1.5 V and the sensor is floating.

In Fig. 9 (a), the data are extracted from Fig. 8 (a) along the cutline staring from (x,y) = (0,-5) to (0,5). The range covers the whole range of inner circle which possess a radius of 4 μm. We can observe that the current density along the fringe is indeed higher, and that the distribution is quite symmetric so the minimum is located in the center of the inner circle. On the other hand in Fig. 9 (b), we extract the current density of the outer ring along the cutline staring from (x,y) = (0,-17) to (0,-4.5). The range of the outer ring will be (0,-14) to (0,-6) and therefore the chosen range is able to cover it.

There are two things that should be noted from the result of Fig. 9 (b). The first thing is that in the case of a ring-shaped MIS(p) TD which possesses two fringes with different curvature, the majority of the currents still flows along fringes. The second thing is that the current density near the outer fringe of the ring is higher than that of inner fringe and therefore the minimum point no longer locates right in the middle but tends to shift towards the inner fringe.

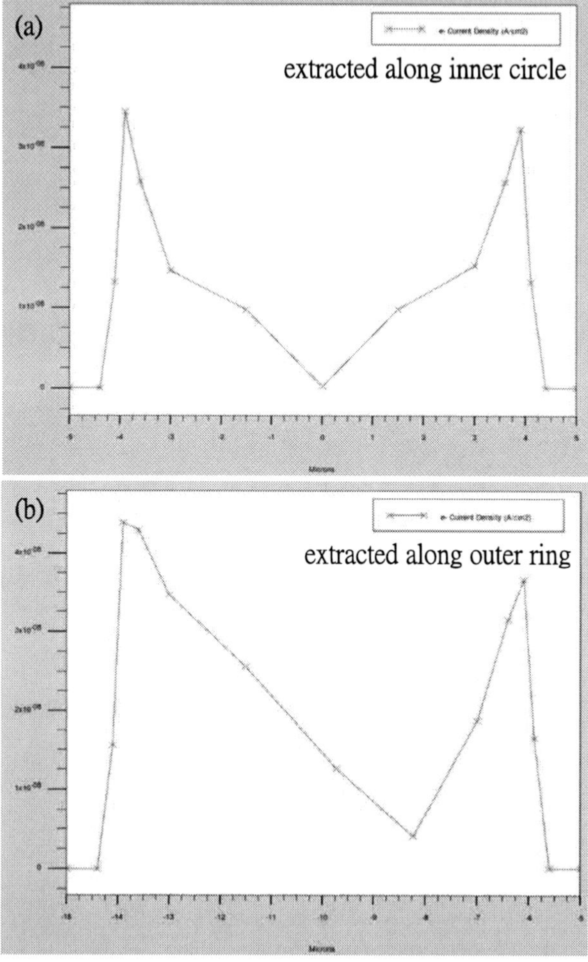

Figure 9. Electron current density extracted along a certain cutline in Fig. 8 under (a) IGOS and (b) ISOG operations. The range of the cutline is chosen to cover the region upon which the 1.5 V bias is added, respectively. The start and end points of the cutline will be (0,-5) to (0,5) in Fig. 8 (a) and (0,-17s) to (0,-4.5) in Fig. 8 (b).

From the above observation we can say that the dominant role in grabbing charges is indeed the outer fringe of the outer ring. This result serves as the strong supporting evidence to our experimental result where the currents of outer rings in set 1 which possess same outer fringe are nearly the same (Fig. 2 (a)) while the currents of set 2 with same inner fringe do not have such characteristic.

Conclusion

In this study, the difference of the coupling ability of the inner circle and outer ring (asymmetric coupling effect) and the unequal ability to collect charges of the outer and inner fringe of a ring-shaped MIS(p) TD are shown from both an experimental and a simulated perspective. It is concluded that the weaker coupling field of the inner circle along with the fact that the majority of the current of the outer ring flow along outer fringe further lead to the consequence in Fig. 4 (a) that the outer sensing currents are much less affected by the inner control gate bias. The findings in this work can help to explain this distinct outcome in a thorough way. Besides, the application of the IGOS situation has been published in our previous paper (4). Further studies on the design of the outer ring-to-inner circle area ratio might be important to optimize the usage of the area while maintaining good performance at the same time.

Acknowledgments

This work is supported by the Ministry of Science and Technology of Taiwan, ROC, under Contract No. MOST 105-2221-E-002-180-MY3 and MOST 107-2622-8-002-018.

References

1. Y. K. Lin, J. G. Hwu, *IEEE Trans. Electron Devices* **61**, 3217–3222 (2014).
2. Y. K. Lin, L. Lin and J. G. Hwu, *ECS J. Solid State Sci. Technol.,* **6**, 132–135 (2014).
3. W. T. Hou and J. G. Hwu, *ECS J. Solid State Sci. Technol.,* **6**, 143–147 (2017).
4. Y. H. Chen, J. G. Hwu, *IEEE Trans. Electron Devices* **65**, 4910 - 4915 (2018).

ECS Transactions, 89 (3) 133-136 (2019)
10.1149/08903.0133ecst ©The Electrochemical Society

Effect of Silver Nanoparticles on the Electrical Characteristics of Oxide/Semiconductor Heterojunctions

A. Rezk[a], Y. Abbas[a], I. Saadat[b], A. Nayfeh[b], M. Rezeq[a,b]

[a] Department of Physics, Khalifa University, Abu Dhabi 127788, UAE
[b] Electrical and Computer Engineering Department, Khalifa University, Abu Dhabi 127788, UAE

The localized electrical effect of nanoparticles present in an oxide/semiconductor hetero-junction is investigated by means of Conductive Atomic Force Microscopy (CAFM) and Scanning Kelvin Probe Microscopy (SKPM). This study demonstrates the capability of the sandwiched NPs to modulate the oxide/semiconductor junction in both forward and reverse directions, thus enabling a distinct nano-scale heterojunction. This research, in turn, will contribute to the miniaturization of such NPs based devices in novel applications.

Introduction

The electrostatic properties at the nanoscale of hetero-junctions in specific environments can be potentially useful in novel electrical and optical applications that bulk materials and devices cannot provide. For instance, nanoparticles (NPs) based structures have drawn a lot of attention due to their unique electrical characteristics and applications in nano-devices [1-4]. A complete understanding of the exact physics underlying heterogeneous junctions between metallic nano-particles or islands and thin films of high-k dielectrics, is yet to be formulated. Effects such as Maxwell–Wagner polarization and interfacial charge are yet to be investigated on such junctions [5,6]. Embedding different kinds of NPs in thin dielectric oxides has shown to play a key role in nano-scale heterojunction devices. NPs based heterojunction devices can be suitable candidates for the next generation technology, especially with the industry's aggressive scaling requirements.

Experimental

Towards this purpose, the effect of silver (Ag) nanoparticles (around 40 nm in diameter) on the electrical characteristics of oxide/semiconductor heterojunctions is investigated by using the Conductive Atomic Force Microscopy (CAFM) and Scanning Kelvin Probe Microscopy (SKPM). The AFM allows us to bias an individual NP to investigate and map its electrical characteristics. For this purpose, the uniform and individual dispersion of these nanoparticles is optimized by employing the appropriate sonication, dispersion methods and thermal treatment. Spin coating was done at 2500 RPM for 45 s after dropping a 0.5 ml on the samples while spinning for 10 s at 500 RPM. Drop casting was conducted using 1 ml of Ag NPs suspended in a citrate buffer with ~ 5.5×10^{13} particles/ml concentration. The NPs are dispersed directly on an n-type Si substrate (5 Ohm.cm) as well as sandwiched between two layers of Al_2O_3 on the Si substrate. The Al_2O_3 films were deposited with Atomic Layer Deposition (ALD) at 200 °C and 20 mTorr, in an Oxford Instrument reactor using trimethyl-aluminum (TMAl, $Al(CH_3)_3$) as metal organic precursor, with remote O_2 plasma as a co-reactant. The size and dispersion of the nanoparticles on both substrates are determined by means of AFM topography and SEM micrographs. The samples were heated at 150 °C on a hotplate to assure the removal of all surfactants. Using these optimized dispersion conditions, we prepared two sets of structures consisting of the following two stacks [Ag-NPs/Si] and [Al_2O_3 (3 nm)/Ag-NPs/Al_2O_3 (1 nm)/Si]. Two conformal tunneling and blocking oxide layer of Al_2O_3 are deposited by using atomic layer deposition at 200 °C, with thicknesses of 1 nm and 3 nm, respectively. For comparison of the electrical characteristics, two sets of control structures are prepared consisting of an HF-cleaned bare Si substrate and a Si substrate coated with 4 nm Al_2O_3. The effect of these individual NPs on the electrical characteristics is determined and compared against these control structures. For this purpose, we initially carried out the topography scans of the above-mentioned structures in tapping mode of AFM, then executed the electrical characterization (I-V) by softly engaging an individual and representative NP. Both CAFM and SKPM were done using a conductive n^+-silicon Au-coated (0.01-0.02 Ω.cm) cantilever tip with a force constant between 1.2-29 N/m, a resonance frequency between 76-263 kHz and a tip radius inspected under SEM to be typically 33 nm.

Results and Discussion

<u>Conductive Atomic Force Microscopy (CAFM)</u>

The electrical response for both control structures during the application of voltage sweeps of +5 V ~ 0 ~ -5 V, showed the rectifying behavior at the sample bias. Interestingly, the electrical characteristics of the NPs sandwiched between the two oxides appear to have enhanced current (reversed rectification) at positive bias during the application of the same voltage sweep as shown in Figure 1 (the inset shows an AFM image of the NPs embedded between two layers of oxide). This indicates that the NPs play a key role in the current transport. The NPs can form tunneling paths connecting the Si substrate to the AFM tip, and thus a large current can be expected due to the local field enhancement between the NP and the substrate interface [1-4], which also enhances further by decreasing the tip size [7-9]. The current observed for the [Al_2O_3/Ag-NPs/Al_2O_3/Si] stack is more significant as compared to the control structures and thus the sandwiched structure seems the best candidate for heterojunction-based applications. At highly positive voltage sweeps (~7 V), the same effect behavior holds but can be shadowed by the large current generated under a highly negative sweep.

The steeper I-V plot can be attributed to the interfacial polarization connected to the free charges generated at the interfaces between the metallic NPs and both dielectric layers. The effect can be explained by the so-called Maxwell-Wagner theory on heterogeneous dielectric media that have a discontinuity in the oscillating polarizations at the interfaces of two media having different dielectric constant and electrical conductivity [10,11].

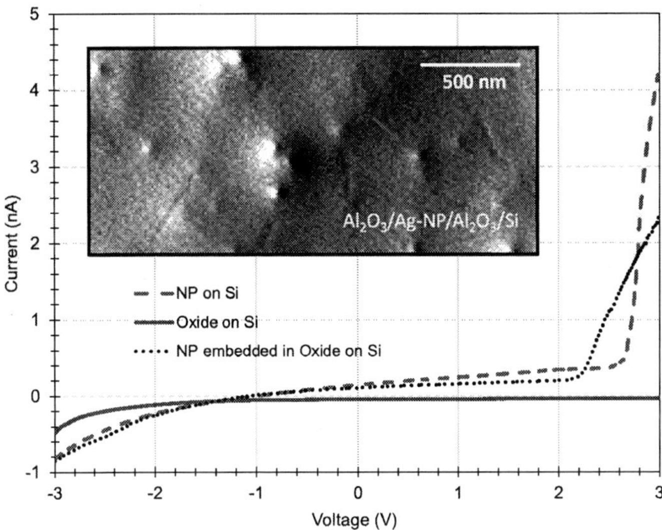

Figure 1. I-V characteristics of the Ag-NP/n-type Si junction (dashed red line), the Al_2O_3/n-type Si junction (solid blue line) and the Al_2O_3/Ag-NP/Al_2O_3/n-type Si junction (dotted black line). The inset shows an AFM image of the NPs embedded between two layers of Al_2O_3.

<u>Scanning Kelvin Probe Microscopy (SKPM)</u>

To have more insight on these results we further probed the surface potential of the [Al_2O_3/Ag-NPs/Al_2O_3/Si] stack using Scanning Kelvin Probe Microscopy (SKPM). From SKPM analysis, it is observed that the potential is dipping at the location of nanoparticles as depicted by Figure 2. This potential drop indicates that the electrons are more easily attracted towards the Si substrate during the positive voltage sweep.

Conclusion

We have demonstrated that the presence of Ag nanoparticles in an oxide/semiconductor hetero-junction enhances the localized electrical field along with the surface potential of thin film stacks. This results in the enhancement of the tunneling current through the oxide layers under a reverse bias, which would highly stimulate the development of NPs based efficient nano-devices.

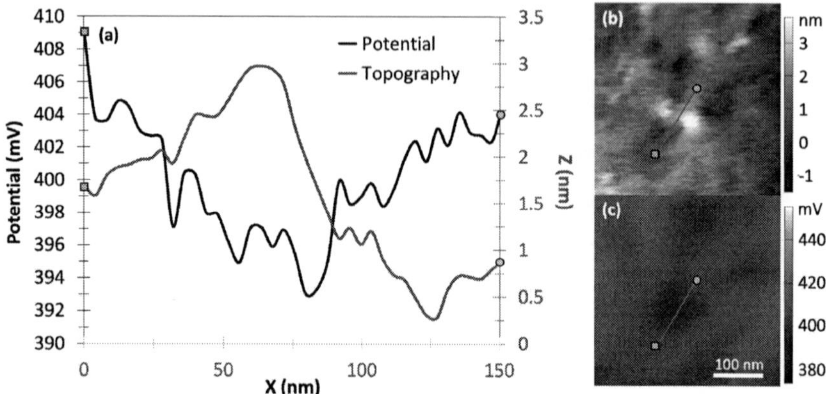

Figure 2. (a) Cross-sectional scans along a ~40 nm diameter Ag-NP embedded between the two layers of oxide showing the height profile (solid red curve) and the measured surface potential across the same scan vector (solid black curve); (b) and (c) show the AFM topographical height and the corresponding mapped surface potential for the plot in (a). The red lines in (b) and (c) show the trajectories of the scanning probes.

References

1. M. Rezeq, A. Ali, S. P. Patole, K. Eledlebi, R. K. Dey and B. Cui, *AIP Advances*, **8**, 055122 (2018).
2. M. Rezeq, K. Eledlebi, M. Ismail, R. K. Dey and B. Cui, Theoretical and experimental investigations of nano-Schottky contacts, *Journal of Applied Physics*, **120**, 044302 (2016).
3. C. S. Pathak, M. Garg, J. P. Singh and R. Singh, *Semicond. Sci. Technol.* **33**, 055006 (2018).
4. E. Guo, Z. Zeng, X. Shi, X. Long and X. Wang, *Langmuir*, **32** (41), 10589–10596 (2016).
5. J.C. Maxwell, in: *Electricity and Magnetism*, Vol. 1, Clarendon Press, Oxford, 1892, p. 452.
6. R.W. Sillars, *The behavior of polar molecules in solid paraffin wax*, Proc. Roy. Soc. (London) A, **169** (1939) 66–82.
7. M. Rezeq, *Applied Surface Science,* **258**, 1750–1755 (2011).
8. M. Rezeq, *Microelectronic Engineering*, **102**, 2–5 (2013).
9. M. Rezeq, A. Ali and H. Barada, *Applied Surface Science*, **333**, 104–109 (2015)
10. E. Vogel, K. Ahmed, B. Hornung, W. Henson, P. McLarty, G. Lucovsky, J. Hauser and J. Wortman, *IEEE Transactions on Electron Devices*, **45** (6), 1350 (1998).
11. J. Thisayukta, H. Shiraki, Y. Sakai, T. Masumi, S. Kundu, Y. Shiraishi and S. Kobayashi, *Japanese Journal of Applied Physics*, **43**(8A), 5430–5434 (2004).

ECS Transactions, 89 (3) 137-153 (2019)
10.1149/08903.0137ecst ©The Electrochemical Society

Laser Thermal Annealing for Low Thermal Budget Applications:
from Contact Formation to Material Modification

K. Huet[a], T. Tabata[a], J. Aubin[a], F. Rozé[a], L. Thuries[a], S. Halty[a], B. Curvers[a],
F. Mazzamuto[a], J. Liu[b] and Y. Mori[c]

[a] SCREEN-LASSE, 92230 Gennevilliers, France
[b] SCREEN SPE, Albany, NY 12206, USA
[c] SCREEN Semiconductor Solutions, Co., Ltd., Hikone, Shiga 522-0292, Japan

Pulsed laser thermal annealing at the nanosecond timescale was successfully introduced into semiconductor device manufacturing in the late 2000s for high volume manufacturing of sensitive 3D architectures such as vertical Si-based Insulated Gate Bipolar Transistors, SiC-based vertical power diodes and backside illuminated CMOS imaging sensors. It is now on the verge of being integrated in key annealing process steps in next generation CMOS and memory devices manufacturing. This invited paper presents recent experimental and simulation work involving sub-μs laser annealing for a wide range of applications, from contact formation to material modification and 3D sequential integration.

Introduction

In the history of semiconductor device fabrication, annealing is one of the critical process steps used to achieve targeted device performance. During the period of traditional Moore's law scaling, the thermal budget of key annealing steps, typically for junction, gate and contact formation modules, had to be continuously reduced to find the best tradeoff between activation, controlled species diffusion, crystallization quality and defects curing. As depicted in Figure 1, the idea is to reduce the annealing timescale to achieve higher anneal temperatures, thus allowing better control of the thermal budget and improving material properties. Lowering the time scale and, consequently, the spatial scale, comes with its own set of challenges, especially concerning the temperature variation scale (1), from wafer to wafer (batch furnace) to within wafer (single wafer RTP and spike) to within die (millisecond anneal) and finally to within layout (sub-μs anneal) process variability. While many of these challenges have been overcome over the years for millisecond and longer anneal schemes by improving the equipment (e.g. spatial distribution of backside heating for spike), adjusting the process flow (e.g. combination of RTP or spike with millisecond anneal) or adapting the layout design (2, 3), they are still being actively worked on for sub-μs annealing.

Pulsed Laser Thermal Annealing (LTA) of semiconductor materials at sub-microsecond timescales has been investigated for many years (4 and references therein) with a main focus on shallow junction formation in the melting regime. For CMOS manufacturing, the potential benefits such as highly localized dopant activation above the solid phase equilibrium limit (5), near-perfect liquid-phase induced recrystallization (4) were hindered by integration issues such as deactivation from the thermal budget from following steps in the flow (6) and pattern variations at transistor level. However, as a preliminary step demonstrating the high-volume manufacturing readiness of the technology, pulsed laser annealing was adopted in niche markets such as power device (silicon and silicon carbide based) and backside imaging sensors. The common feature for adoption are three-fold: the layout pattern is not directly visible to the laser effect (buried deeper than the thermal diffusion length); the devices are intrinsically 3D and require the thermal budget to be selectively localized near the surface, while the buried layers are kept below a given temperature; the overall thermal budget of process steps following laser anneal is low enough to keep its outstanding benefits.

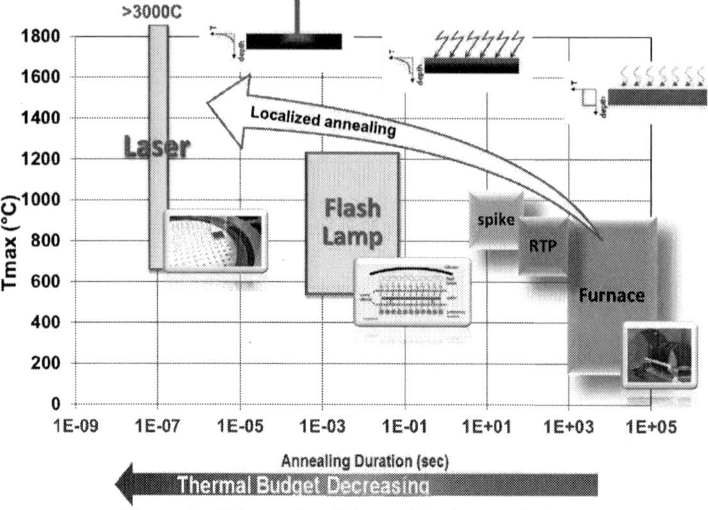

Figure 1. Schematic of thermal budget evolution.

The introduction of new materials (especially low temperature ones), new process schemes and architectures may open the path to a successful integration of pulsed laser anneal into mainstream CMOS and memory manufacturing (4, 7). In this paper, after a brief description of laser annealing process simulation and associated physics, recent work on its most promising applications will be reviewed. In the first part we will focus on its application to the traditional CMOS manufacturing flow, while in the second part we will detail its application to 3D architectures.

Nanosecond Laser Annealing Simulation

One of the key items which has driven the adoption of many breakthrough processes is the ability to understand and predict them through accurate numerical simulations, especially challenging in the case of laser annealing involving phase change (8, 9, 10). We have recently developed a computational tool dedicated to the simulation of LTA process for 1D, 2D and 3D structures as a user-friendly TCAD package (12), called LASSE Innovation and Application Booster (LIAB) and based on a finite element method solution of partial differential equations. It solves self-consistently the heat equation coupled to the time-harmonic solution of Maxwell equations (laser light coupling), phase field and species diffusion, including temperature dependency of material parameters, phase change and alloy fraction. The core model equations [1] through [5] reported below, more detailed in refs. (8-11), are integrated in a modular fashion, allowing for simple implementation of new physics and features, with a user-friendly interface for simulation structure and equipment parameters settings, batch parametric splits submission and tracking, efficient post processing and analysis, and automated report generation. As a TCAD tool, it can be used to estimate the optimal LTA process insertion point in the manufacturing flow for a given target and to evaluate the impact of process parameters (e.g. laser energy density, pulse duration, assist heating temperature) and structure characteristics (e.g. shape, stacks, materials, doping concentration) on the annealing dynamics.

$$\frac{\partial \varphi}{\partial t} = D_\varphi \nabla^2(\varphi) - \frac{\partial F(c, \varphi, \lambda u)}{\partial \varphi} \tag{1}$$

$$\frac{\partial c}{\partial t} = \nabla[D_c(\varphi)\nabla c] - D_c(\varphi)\ln(k_0)\,\nabla[M_2 c(1-c)\varphi(1-\varphi)\nabla\varphi] \tag{2}$$

$$\frac{\partial u}{\partial t} = \rho c_p \frac{\partial T}{\partial t} + [(1-c)L_S + c\Delta L_{S-X}]\frac{\partial h}{\partial t} = \nabla[K(\varphi)\nabla T] + S(r,t) \tag{3}$$

$$S(\boldsymbol{r},t) = \frac{\varepsilon_i}{2\rho}|\boldsymbol{E}_{t-h}|^2 \tag{4}$$

where $\boldsymbol{r}$ is the position, t is the time, φ is the phase (1 for solid, 0 for liquid and -1 for amorphous), T the temperature, c the dopant concentration and u the normalized enthalpy (subscripts S and X are for the pure material and solute, respectively). $F(c, \varphi, \lambda u) = f(\varphi) + \lambda g(\varphi)[u_X(T)c + u_S(T)(1-c)]$ depends on the phase field functions used. F is the Helmoltz free energy functional, f the Helmoltz free energy density, g and h are related functions chosen to obtain the adequate shape of F for the phase field formulation in the sharp interface limit. $S(\boldsymbol{r},t)$ is the heat source from the laser light coupling with the irradiated structure, ε_i is the imaginary part of the temperature dependent complex dielectric constant of the material and $\boldsymbol{E}_{t-h}$ is the time harmonic electric field (from the solution of the corresponding Maxwell equations). The temperature and phase dependent material and species properties used here are the latent heat of the pure material L_S, ΔL_{S-X} its change due to the solute, the density ρ, the specific heat c_p, the thermal the conductivity K, the equilibrium segregation coefficient k_0, the dopant diffusivity D_c and M_2 is a coefficient related to the solute mobility where $D_c = M_2 V(T)/(k_B T)$. λ is a constant related to the interface kinetics and especially the interface velocity in the form:

$$V(T) \,\alpha\, \exp\left[\frac{-E_a}{k_B T}\right]\left[1 - \exp\left[\left(\frac{\rho L}{k_B N}\right) \times \left(\frac{1}{T_M} - \frac{1}{T}\right)\right]\right] \tag{5}$$

where E_a is the activation energy for the transition of the atoms from the liquid to the solid phase, k_B the Boltzmann constant, N the atomic density and T_M the melting temperature of the material. With this formulation, the evolution of the melting front is naturally driven by the thermal field, including the influence of undercooling and interface curvature. Relevant examples of use cases will be detailed in this paper.

Application to CMOS Contact Module

As stated in the introduction, LTA is potentially better suited for applications involving lower temperature materials. Moreover, in the traditional CMOS manufacturing flow, the lower thermal budget steps are by nature used near the end of the flow in modules from the Middle End of Line (MEOL) and Back End of Line (BEOL). In this part of the review we will detail recent activity in the MEOL for contact module improvement, from blanket studies to device application, focused on p-type contact formation.

<u>Material Studies on Blanket Structures</u>

Silicon-Germanium is now commonly used for the source/drain region and even the channel of FinFET type PMOS devices. However, achieving low contact resistivity on such layers remains a challenge, especially due to the low solubility limit of dopants in SiGe. From recent work, a promising pathway to improvement relies on engineering the Ge fraction profile as well as the dopant distribution in the SiGe layer. In (13), 30 nm thick undoped $Si_{0.8}Ge_{0.2}$ epitaxial layers were laser annealed in various conditions and extensively characterized by haze, Atomic Force Microscopy (AFM), X-Ray Diffraction (XRD) for strain, Time of Flight Secondary Ion Mass Spectroscopy (ToF-SIMS) and Transmission Electron Microscopy (TEM) coupled to Energy Dispersive X-Ray (EDX).

Depending on the laser Energy Density (ED), 4 main regimes could be observed (from low to high ED): the sub-melt (I), partial melt (II), full melt (III) and melt beyond SiGe/Si interface (IV). These regimes correspond to different levels of strain relaxation in the layer as reproduced in Figure 2, from strained (I) to partially relaxed (II) then back to strained (III and IV). It was found that the full melt regime (III) would be the most interesting for application to the contact module. In this condition, the SiGe layer is pseudomorphic and shows perfect crystal quality (from TEM) with very limited relaxation (~2 % from XRD).

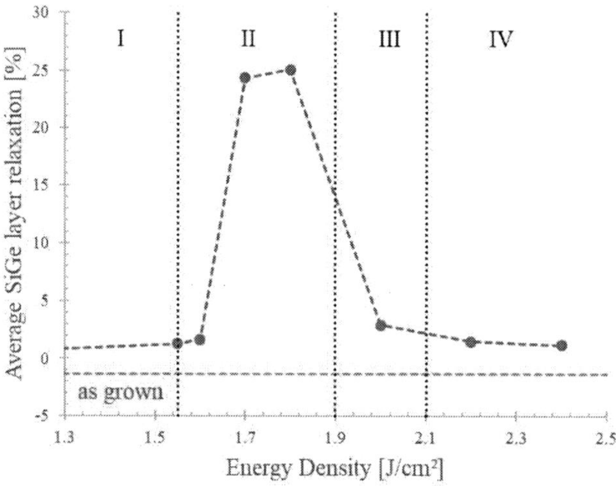

Figure 2. Macroscopic degree of strain relaxation in the laser annealed SiGe layer, extracted from XRD Reciprocal Space Mapping; extracted from (13).

As it can be seen on the ToF SIMS reported in Figure 3 (reproduced from data reported in (13)), the Ge profile was strongly impacted by the melting process. In the partial and full melt profiles, the SiGe/Ge interface remains while the Ge is redistributed within the molten layer with a higher concentration near surface and some depletion near the maximum melt depth which is typical of a segregation phenomenon.

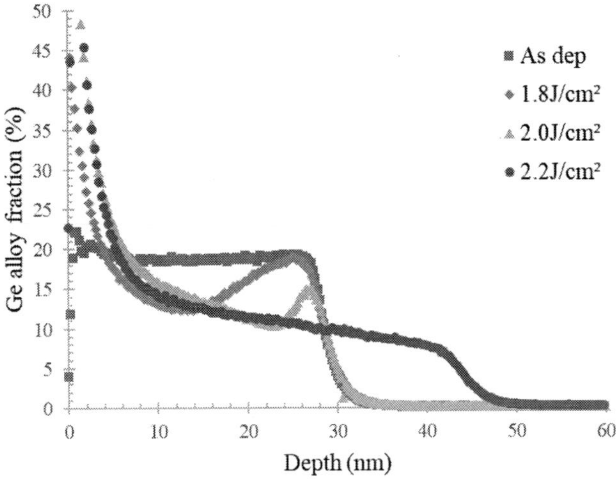

Figure 3. ToF-SIMS germanium atomic concentration depth profiles; reproduced from (13).

This kind of strained Ge-rich surface is of particular interest for Ti-Ge contact resistance reduction for PMOS devices as demonstrated in (14). The improvement was shown to be related to the Fermi level pinning (FLP) moving closer to the valence band and strained-induced reduction of SiGe bandgap. To further enhance such effects, the use of group III elements with low segregation coefficient (k) such as Gallium for doping may be beneficial. In (15), partially relaxed epitaxial $Si_{0.5}Ge_{0.5}$ layers were implanted with a high dose Ga (26 keV, 10^{16} at/cm²), thus amorphizing most of the SiGe layer. Upon laser annealing, samples were characterized by SIMS, Electrochemical capacitance-voltage profiling (ECVP), four-point probe sheet resistance and TEM. Regimes similar to (13) were observed with the notable difference of the presence of an explosive crystallization (EC) phenomenon which lead to conversion of the amorphous layer into nanocrystalline poly-SiGe and a uniform redistribution of the dopants at the initial stages of the process, as already observed in previous studies (11).

As shown in the inset of Figure 4, for the laser condition near full melt of the SiGe layer, and in agreement with our simulation predictions, Ge segregation during the melting process lead to a graded Ge profile with a Ge-rich region near surface and a more Si-rich region near the maximum melt depth. Interestingly, the Ga concentration profile also shows strong surface segregation. Simulation shows that such profile can only be reproduced by considering a liquid solubility about 10 times higher than the solid one. ECVP confirmed that in the shallow melt case, Ga could indeed be activated about 6 times over the solid solubility limit. Together with the Ge-rich surface, this can be highly beneficial for moving the FLP closer to the valence band and further improve contact resistivity.

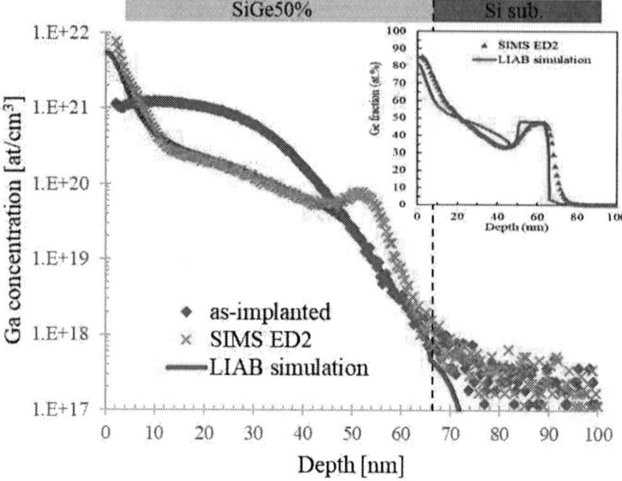

Figure 4. Ga SIMS profiles as implanted (solid diamonds) and after LTA for near full melt conditions (crosses), superposed to LIAB simulation (solid line). Inset: corresponding Ge fraction SIMS profile (solid triangles) superposed to simulation (solid line); from (15).

By a finer analysis of simulation results, some degree of undercooling (50 °C range) was determined during the solidification stage, which may shed some light on the dynamics of the process. As discussed in (16, 17) undercooling may be one of the driving forces for the precipitation of inactive dopants near the surface through the liquid/solid interface during the solidification stage. Through the reduction of the solidification velocity (and therefore the amount of dopants involved in precipitation) by optimizing the implant and laser conditions (pulse duration, assisted heat), it should be possible to maximize the activation rate beyond the solid solubility limit in the region of interest, near the metal/semiconductor interface.

<u>Application to Patterned Structures</u>

In order to assess the applicability of these findings to realistic structures and obtain meaningful electrical figures of merit, contact resistivity was extracted by Multi-Ring Circular Transmission Line Method (MR-CTLM) using a similar process flow, shown in Figure 5(a) (18, 19, 20). This kind of approach, detailed in Figure 5(b), is specifically optimized to allow an accurate determination of contact resistivities below 10^{-8} $\Omega.cm^2$ with an error of $2 \cdot 10^{-10}$ $\Omega.cm^2$ using TEM for the extraction of the spacing values of the MR-CTLM structure (18).

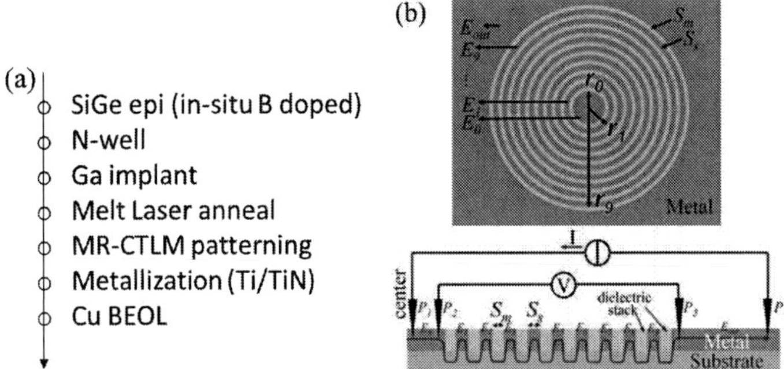

Figure 5. (a) Simplified MR-CTLM process flow, reproduced from (18); (b) Schematic layout of MR-CTLM structure from (20).

In the work reported in (18), a high Ge content SiGe layer (~60 % Ge) was epitaxially grown on a lightly doped Si substrate and implanted with high dose Ga ($>10^{15}$ at/cm²) to a shallow depth (< 10 nm) before melt laser anneal and subsequent patterning and metallization steps, with the notable use of Ti in the contact module. Various laser conditions (fluence, number of pulses, pulse duration) were applied to find the optimized conditions to minimize contact resistivity down to a record low value of $1.3 \cdot 10^{-10}$ Ω.cm².

Ge and Ga distributions, measured by Electron Diffraction Spectroscopy (EDS) and schematically reproduced in Figure 6(a) show that both Ge and Ga strongly segregate near the Ti/SiGe interface, confirming the findings from the blanket experiments detailed previously. Moreover, as shown in Figure 6(b), the contact resistivity reduction was found to be well correlated with the Ga concentration at the Ti/SiGe interface. Although EDS does not give the active content of Ga, an increase in active concentration N can be expected. This seems to further confirm that at the Schottky-barrier contact formed by MLA segregation process, the field-emission dominates carrier transport.

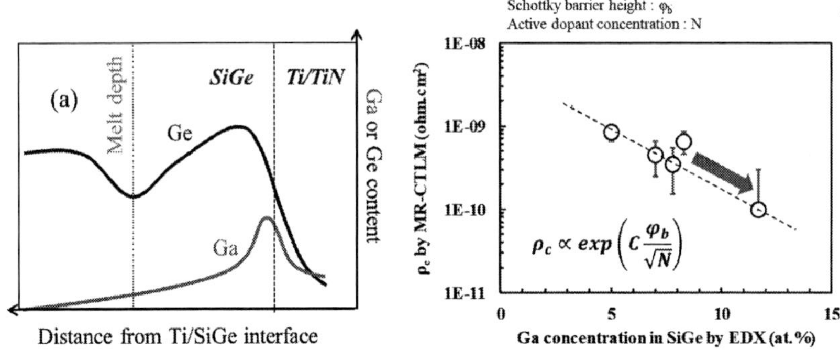

Figure 6. Left: schematic Ge and Ga content profiles from EDS, Ga-content is scaled for better readability; right: Contact resistivity vs. Ga-fraction at the Ti/SiGe interface (18).

<u>Application to FinFET Device Structure</u>

A very similar approach was then applied to p-type finFET device (21). The simplified process flow is detailed in Figure 7 for conventional p-type finFET devices with replacement metal gate (RMG) scheme and in-situ Boron doped SiGe epi source/drain. One of the key points for a successful integration is to perform implant and melt laser anneal steps after contact opening. In this case, the exposed in-situ doped SiGe epi area was first preamorphized by Ge implant (Ge-PAI) and implanted either by traditional beamline or plasma implantation (PLAD) technique before laser annealing.

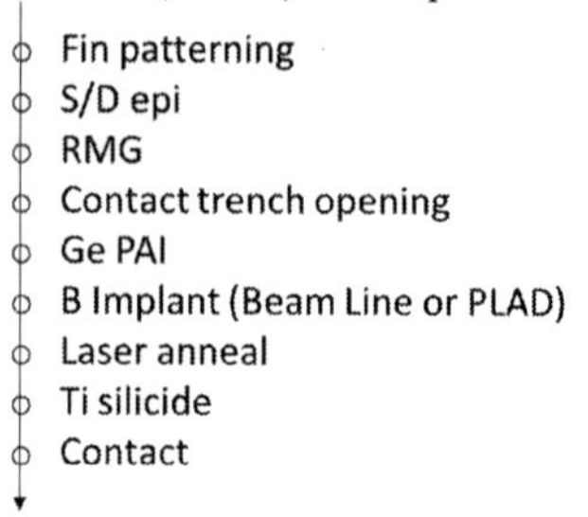

- Fin patterning
- S/D epi
- RMG
- Contact trench opening
- Ge PAI
- B Implant (Beam Line or PLAD)
- Laser anneal
- Ti silicide
- Contact

Figure 7. Simplified FinFET device process flow with laser melting; reproduced from (21).

Contact resistivity comparison between the non-laser annealed and laser annealed cases reproduced in Figure 8 clearly demonstrate the intrinsic benefit of the melt laser annealing approach and its applicability to finFET devices, with up to 50 % improvement observed (24 % by melt laser anneal alone). This translates to a corresponding on-current benefit up to about 8% at comparable off-current, with well-behaved current-voltage characteristics, showing good control of the short channel effects. Further improvement was obtained by applying an additional millisecond non-melt laser annealing in the case of beamline implant, but the best performance was achieved by PLAD and melt laser anneal alone. This can be explained by the good conformality and high dose of the plasma doping approach, combined with the shallow diffusion and super-activation benefits of the melt laser process.

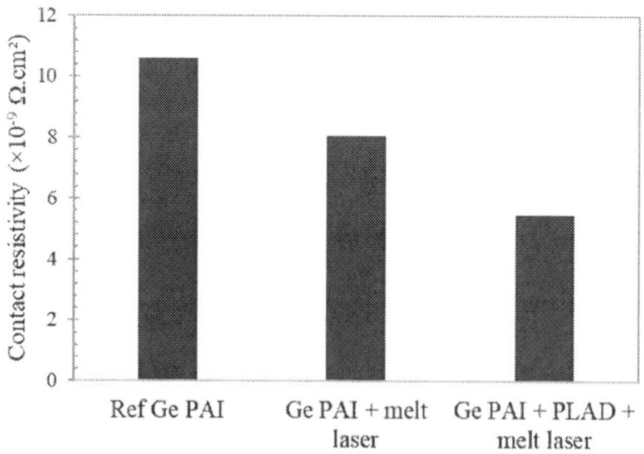

Figure 8: P-type FinFET contact resistivity with and without melt laser anneal; from (21).

Application to Nanowire (NW) Device Structure

The integration of LTA process was further confirmed on actual devices in recent work (22) on lateral nanowire Ge-channel based n-type transistors, one of the most promising architectures for the coming technology nodes. This time, the Source/Drain contact regions are formed using an implant-free approach by optimized multi-layer epitaxy with heavy in-situ doping above solid solubility limit. The dopants were then activated using LTA before the metallization step. The implant-free approach avoids the usual drawbacks of ion implantation such as end-of-range defects, which would potentially require higher ED (deeper melt) to fully recover, with the associated risk of too much in-depth diffusion in the contact fins. In (22), this scheme was first applied to MR-CTLM structures to evaluate its potential for contact resistivity improvement. By optimizing the epitaxy and laser conditions, very low resistivities, down to $1.6 \cdot 10^{-9}$ $\Omega.cm^2$, could be achieved, setting a record for contacts on n-Ge at the time of publication. The optimized process was then successfully integrated in the Ge single nanowire transistor process flow. As a result, a 9 nm diameter NW, 70 nm gate length with scaled contacts and well-behaved transfer characteristics was demonstrated, setting a first for a functional scaled n-type Ge-based nanowire field effect transistor manufactured in a high-volume manufacturing CMOS compatible process flow.

3D simulations of a similarly scaled structure were performed using LIAB to understand the specifics of LTA process on such kind of device and help design optimization. A schematic of the simulated 3D structure is shown in Figure 9. Material optical and thermal properties were taken from available literature (8, 11) and a reduced Ge thermal conductivity compared to the bulk case was considered in the Source/Drain, fin and NW regions to account for size effects on thermal properties (23, 24).

Figure 9. 3D Ge Nanowire schematic of simulated structure.

Heat source distribution calculations (not shown), representative of the local laser light coupling efficiency, demonstrate that, thanks to the very high absorption efficiency at LTA wavelength, most of the coupling occurs in the surface region layers and only a small fraction is absorbed in the gate and channel regions. Heat generated in this area can diffuse to the regions of interest for dopant activation during the annealing process. Figure 10 shows the maps of the maximum temperature reached in the structure during the whole

LTA process duration for a sub-melt regime condition, on cross section planes along the NW direction (left) and along the gate direction (right). As it can be seen, the high temperature region is located near the surface, in the gate and in the NW. In the deeper regions, the temperature drops very quickly due to the high thermal conductivity of the underlying Ge epi layer. The time spent at high temperature extracted from the simulated temperature transients (not shown) is only a few 100ns, which enables very localized and limited diffusion of the dopant species.

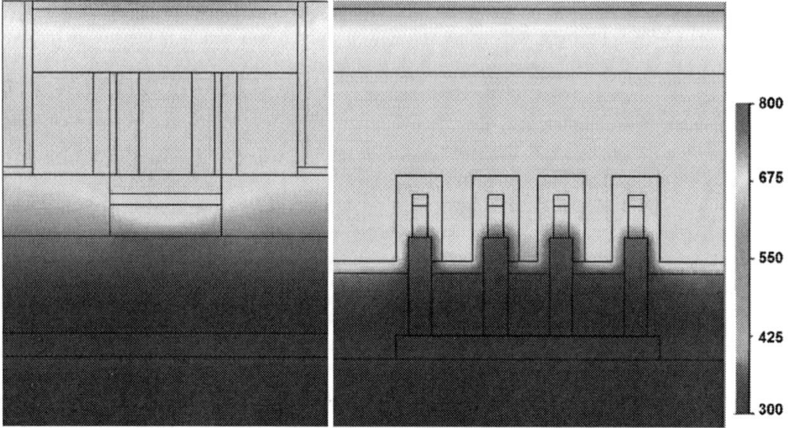

Figure 10. Maximum temperature maps (in °C) for LTA in submelt condition, in cross section planes along fin/NW direction (left) and along gate direction (right).

Finally, a relatively narrow process window width of about 5 % in terms of ED could be determined, defined by the energy threshold of surface melting for the fin or channel melting for the lower and higher limit of the energy density range, respectively. Other simulations (not shown) helped suggest some improvement pathways to enlarge this process window by optimizing some key structure parameters.

While this review focuses on p-type architectures, it should be noted that sub-10^{-9} $\Omega.cm^2$ contact resistivity was also demonstrated on n-type contact chain structures using melt laser anneal (25). More recent work yet to be published investigates process flow optimization paths to enable integration of melt laser anneal in both n- and p-type devices in the same die layout, which will enable its adoption in high-volume logic devices manufacturing.

Application to advanced interconnects

Back end of line is another area of application which is seeing a pressing need of novel solutions to challenges posed by the shrinking dimensions of contact vias and conductive lines and the ever-growing density of interconnects. In this field, material modification involving efficient low thermal budget is critical. In this part, we will detail some of the recent work involving melt laser anneal of copper interconnects for improved microstructure.

<u>Copper Interconnects</u>

BEOL contact resistance is expected to become the bottleneck for circuit performance in future technology nodes. Cu line resistance is known to increase dramatically for narrow interconnect lines as linewidths become smaller than the electron mean free path in copper. As first introduced in (26), melt laser anneal is a promising path to improving copper microstructure, electrical properties and reliability performance while keeping the thermal budget low enough to avoid damaging the fragile surrounding interfaces and low-k dielectrics. In (1), large grains and ~25 % line resistance improvement were shown for linewidths down to 40 nm. The improvement is higher for smaller linewidth and is believed to be due not only to grain size enlargement, but also to a reduction in interface scattering from impurities.

This result was confirmed more recently (27) using a 14 nm node finFET technology. In this work, M1 interconnects were formed using the process flowchart described by Figure 11.

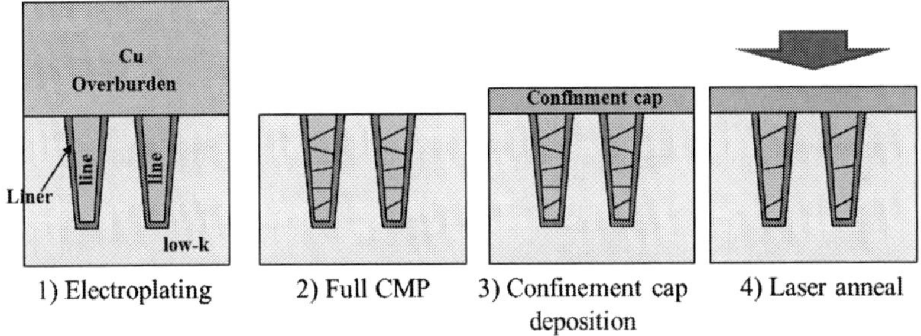

1) Electroplating   2) Full CMP   3) Confinement cap   4) Laser anneal
deposition

Figure 11. Process flow for M1 interconnects with melt laser anneal; adapted from (27).

The key optimization point in the process flow was the removal of the electroplated Cu overburden by Chemical-Mechanical Planarization (CMP) followed by deposition of a back-end compatible low temperature confinement layer before applying melt laser anneal. The laser process parameters (especially the melt laser peak temperature and wafer base temperature) were then optimized to find the best tradeoff between line resistance improvement (up to +35 %) and capacitance degradation (~9 %, likely due some impact on low-k material) for an overall RC delay improvement of ~15 %. The resulting improvement in final device performance after completing the full 11 metal layers was significant (2 to 5 %). Combined with excellent yield characteristics, the robustness of the approach is clearly demonstrated and hints that this could be further improved by applying the laser anneal to more interconnect layers in future technology nodes, in which linewidths at higher metal levels will also get smaller.

As stated in the introduction, one of the main challenges for laser anneal integration is to overcome the variability induced by the pattern differences within a die. As an example, using a structure similar to the one described above (with confinement cap), we simulated the impact of varying Cu line pitch (keeping the linewidth constant) on surface temperature for a given laser condition corresponding to a temperature just below the melting threshold for the 45 nm pitch case. As it can be seen on Figure 12, the pattern effect is quite significant in this case, with up to 300 °C difference between the 90 nm and 45 nm pitch cases. This is essentially due to the difference in laser light coupling to structures with dimensions below the laser wavelength.

These differences can be mitigated by adjusting the process flow (e.g. keeping some Cu overburden to uniformize the laser coupling as initially suggested in (26)) and the laser process parameters such as assisted heating to minimize the impact of the coupling differences (28).

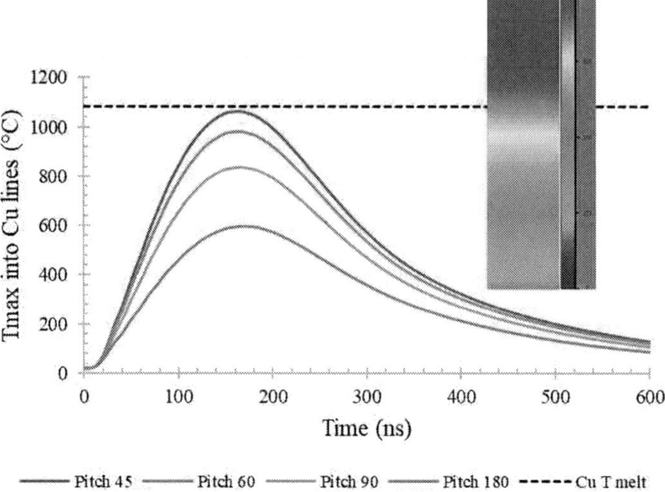

Figure 12. Simulated temperature in Cu lines vs time for different Cu line pitch for a given laser condition near melt threshold. Inset: Temperature distribution in the structure.

Application to 3D Architectures

For 3D oriented devices, LTA can be applied as the low thermal budget solution for material modification, namely polysilicon crystallization. In this part of the review we will detail recent activity on poly gate crystallization for 3D sequential integration and polysilicon plug formation for memory.

<u>3D Sequential Integration</u>

The use of melt laser annealing for large polysilicon grain formation has been studied for decades and successfully used in production for flat panel display applications (29). In parallel, 3D sequential integration scheme, schematized on Figure 14 has been developed as an alternative approach to TSV-based 3D integration (30) with the major challenge being to realize a high-performance transistor at the top level without degrading the electrical characteristics of the bottom one and keeping the thermal budget of the BEOL interlayer below 500 °C, as illustrated in Figure 13.

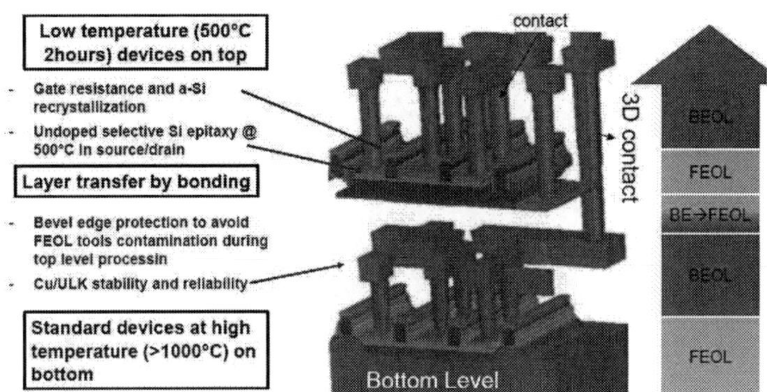

Figure 14. 3D sequential CoolCube™ integration and remaining process issues; reproduced from (32).

These last years, the compatibility of LTA for 3D sequential integration, where an active layer is sequentially added on top of a fragile BEOL stack has been demonstrated by simulation and on a real structure (31). More recently (32), LTA has been successfully used for poly gate crystallization of the top layer in a gate-first scheme. The idea was to crystallize and activate dopants of an in-situ doped amorphous layer before the gate patterning steps. After fine tuning of the laser parameters, a high activation rate and full recrystallization of the layer with large grains (~ few hundred nm size range) were obtained, without impacting BEOL integrity. As a result, the melt laser process could be integrated in the gate formation module for 3D sequential CoolCube™. The key point for a successful integration was to address the surface roughness induced by local grain boundaries after crystallization using an optimized CMP step, similarly to (33).

<u>Memory</u>

Polysilicon plug crystallization by LTA, very similar to the grain-filter technique for pattern-controlled localization of polysilicon grain growth (34) was also evaluated for several memory applications, especially vertical 3D NAND channel formation (35, 36) and emerging memory Si-based selector element crystallization (37). In the case of 3D NAND channel formation, a clear improvement of memory performance was observed as a result of increased grain size and interface defects curing (36). However, this achievement is difficult to scale to μm-scale 3D NAND channel lengths currently used in high-volume manufacturing. Nevertheless, these findings may be applicable to overcome other process challenge in DRAM memory Bit Line contact (BLC) and Storage Node Contact (SNC) formation. These contacts are traditionally formed by in-situ doped polysilicon deposition and recess in the contact holes or trenches before metallization. With the continuous downscaling of DRAM array transistor dimensions, the polysilicon deposition filling in those small contact regions may show some voids or seams (38) which may impact contact resistivity and therefore the overall performance of DRAM operation. Figure 15 shows a schematic DRAM array structure highlighting the SNC region contacting the DRAM array to the capacitors. The thermal budget to address these voids has to be limited to avoid diffusion of the dopants into the transistor region or any change in existing metal layer properties.

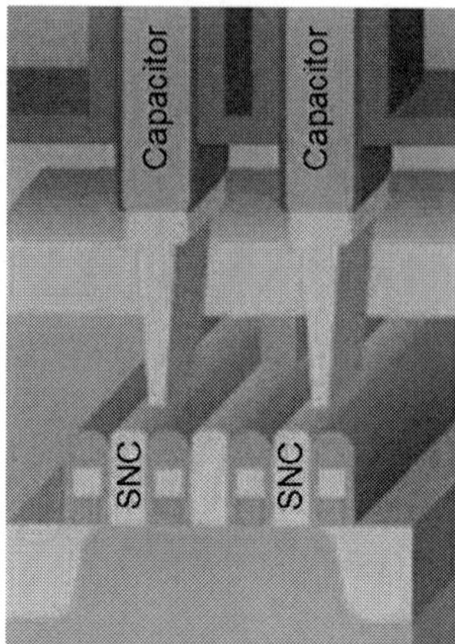

Figure 15. Schematic of a DRAM array structure; reproduced from (39).

Reflow of polysilicon by laser-induced polysilicon melting is an efficient way of removing such voids even in narrow plugs (35), illustrated in Figure 16 for 80 nm plug diameter structures. The reflow depth can even be controlled by tuning the laser parameters (here energy density).

As deposited **As LTA (1.6J/cm2)** **As LTA (2J/cm2)**

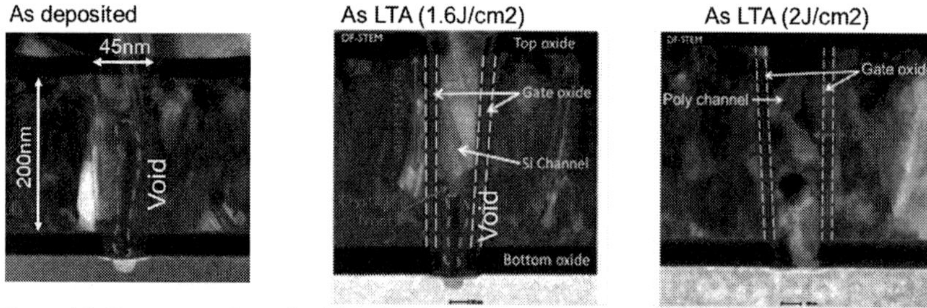

Figure 16. Demonstration of void curing in polySi plugs, from (35). Left: as-deposited plug with the initial void highlighted; center: partial melt condition; right: full melt condition.

From the previous discussions related in this paper, the melt process is compatible with 3D architectures, even in the presence of fragile layers. In the DRAM case, such process could be adapted since the fragile metal layers are insulated from the poly contact by a dielectric layer, which may act as an effective barrier to interdiffusion between contact and active region. To assess the viability of this approach, we have simulated the impact of melt laser anneal on a DRAM-like structure after deposition of the storage node contact polysilicon, before the recess etch step. Structure dimensions were adapted from (40) and a 100 nm Si overburden is assumed. Considering amorphous Si properties for the deposited poly, a process window can be found to selectively melt the SNC contact region without impacting the underlying crystalline part. Figure 17 shows the simulated temperature at the critical interface between the metal layer and the underlying crystal Silicon during the laser process. As it can be seen in the inset, the overall temperature is uniform laterally due to the strong confinement in this direction and shows a vertical temperature gradient of

~100 °C in the structure. The time spent at high temperature is short enough (~300 ns at 1000 °C) to avoid any unwanted interdiffusion of the species. These results show the strong potential of melt laser anneal to address this specific challenge for contact formation in narrow plugs.

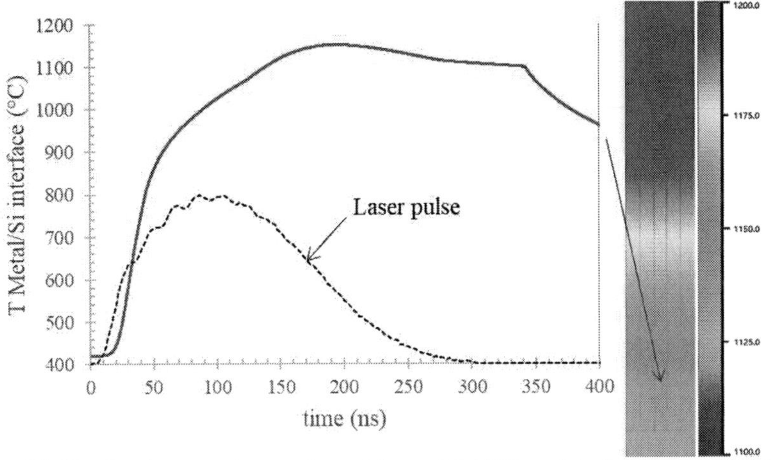

Figure 17. Simulated temperature vs time at the metal / Silicon interface (highlighted by the arrow in the inset), superposed to the laser pulse used in the simulation. Right side: Maximum temperature in the structure during process.

Conclusion

In this paper, we have reviewed the most recent work on integration of Laser Thermal Annealing process, discussing its integration challenges and potential for application in CMOS manufacturing and beyond, supported by simulation performed with our in-house TCAD simulation software dedicated to laser annealing (LIAB). In the traditional CMOS fabrication field, melt laser annealing shows potential in the middle-end of line for the formation of low resistivity contacts thanks to the benefits of localized heating, super-activation and strong surface segregation of dopant and alloy species, especially interesting for p-type contacts. In the back end of line, its applicability to improve copper material properties in narrow lines was demonstrated in a fully integrated flow. Beyond the traditional CMOS field, melt laser anneal could be integrated for 3D sequential stacking of active transistor layers for gate crystallization requiring extremely low thermal budget. Finally, its applicability for the improvement of polysilicon contacts in narrow contact lines and plugs was assessed for DRAM array transistors.

Acknowledgments

The authors would like to acknowledge P. Acosta-Alba, S. Kerdiles P. Batude and L. Dagault from CEA-LETI, Grenoble, France for their collaboration and great insight on SiGe and 3D sequential integration, K. Barla, N. Horiguchi, J. Mitard, S.A. Chew, J.-L. Everaert, M. Schaekers and H. Yu from IMEC, Leuven, Belgium for their collaboration and inspiring discussions on contact module process flow, O. Gluschenkov, D. Sil, T. Nogami and Y. Sulehria from IBM, Albany, USA for fruitful discussions on advanced interconnects.

References

1. O. Gluschenkov and H. Jagannathan, *ECS Trans.* **85** (6), 11-23 (2018).
2. J. C. Scott, O. Gluschenkov, B. Goplen, *et al.*, *IEEE International Symposium on VLSI Technology*, Honolulu, HI, 152-153 (2009).
3. T. Miyashita, T. Kubo, Y. S. Kim, *et al.*, *IEEE International Electron Devices Meeting (IEDM)*, Baltimore, MD, 1-4 (2009).
4. K. Huet, F. Mazzamuto, T. Tabata, *et al.*, *Mat. Sci. Semicon. Proc.*, **62**, 92-102 (2017).
5. C. W. White, B. R. Appleton, and S. R. Wilson, *Laser Annealing of Semiconductors*, edited by J. M. Poate and J. W. Mayer, pp. 111-146, Academic Press (1982).
6. R. Murto, K. Jones, M. Rendon and S. Talwar, *International Conference on Ion Implantation Technology Proceedings. Ion Implantation Technology (IIT)*, Alpbach, 155-158 (2000).
7. H. Besaucèle, A. Adnet, F. Beau, *et al.*, *Proc. SPIE, XXII International Symposium on High Power Laser Systems and Applications*, 110420S (2019).
8. A. La Magna, P. Alippi, V. Privitera, *et al.*, *J. Appl. Phys.*, **95** (9), 4806-4814, (2004).
9. G. Fisicaro and A. La Magna, *J. Comput. Electron.*, **13** (1) (2014) 70-94.
10. S.F. Lombardo, S. Boninelli, F. Cristiano, *et al.*, *Mat. Sci. Semicon. Proc.*, **62**, 80-91 (2017).
11. S. F. Lombardo, S. Boninelli, F. Cristiano, *et al.*, *J. Appl. Phys.*, **123** (10), 105105 (2018).
12. S. F. Lombardo, G. Fisicaro, I. Deretzis, *et al.*, *Appl. Surf. Sci.*, **467**, 666-672 (2019).
13. L. Dagault, P. Acosta-Alba, S. Kerdilès, *et al.*, *ECS Trans.* **86** (7), 29-39 (2018)
14. C.-N. Ni, K.-V. Rao, F. Khaja, *et al.*, *IEEE International Symposium on VLSI Technology*, 1-2 (2016).
15. T. Tabata, J. Aubin, K. Huet, *et al.*, *22nd International Conference on Ion Implantation Technology (IIT)*, 1-4 (2018).
16. T. Tabata, J. Aubin, K. Huet, *et al.*, *3rd Electron Devices Technology and Manufacturing (EDTM) Conference*, to be published in *IEEE J. of Electron Dev. Soc.* (2019).
17. T. Tabata, J. Aubin, K. Huet and F. Mazzamuto, submitted to *J. Appl. Phys.*
18. J.-L. Everaert, M. Schaekers, H. Yu, *et al.*, *IEEE International Symposium on VLSI Technology*, 214-215 (2017).
19. L.-L. Wang, H. Yu, M. Schaekers, J.-L. Everaert, *et al.*, *IEEE International Electron Devices Meeting (IEDM)*, 549-552 (2017).
20. H. Yu, M. Schaekers, T. Schram, *et al.*, *IEEE Electron Dev. Lett.*, **36** (6), 600-602 (2015).

21. C. I. Li, N. Breil, T. Y. Wen, *et al.*, *IEEE Electron Dev. Lett.*, **40** (2), pp. 307-309 (2019).
22. M.J.H. van Dal, G. Vellianitis, G. Doornbos, *et al.*, *IEEE International Electron Devices Meeting (IEDM)*, San Francisco, CA, 21.1.1-4 (2018).
23. Z. Wang and N. Mingo, *Appl. Phys. Lett.* **97**, 101903 (2010).
24. N. Mingo, L. Yang, D. Li and A. Majumdar, *Nano Letters*, **3** (12), 1713–1716 (2003).
25. R. Hung, F. A. Khaja, K. E Hollar, *et al.*, *International Symposium on VLSI Technology, Systems and Application (VLSI-TSA)*, Hsinchu, pp. 1-2 (2018).
26. T. Nogami *et al.*, USA Patent #6,103,624 (2000).
27. R. T. P. Lee, N. Petrov, J. Kassim, *et al.*, *IEEE Symposium on VLSI Technology*, Honolulu, HI, 61-62 (2018).
28. Y. Wang and A. M. Hawryluk. USA Patent #8,309,474 (2012).
29. E. Fogarassy, B. Prévot, S. de Unamuno, *et al.*, *Mat. Res. Soc. Symp. Proc.*685E, D7.1.1-11 (2001).
30. P. Batude, C. Fenouillet-Beranger, L. Pasini, *et al.*, *IEEE International Symposium on VLSI Technology*, Kyoto, T48-T49 (2015).
31. C. Fenouillet-Beranger, P. Batude, L. Brunet, *et al.*, *IEEE SOI-3D-Subthreshold Microelectronics Technology Unified Conference (S3S)*, Burlingame, CA, 1-3 (2016).
32. L. Brunet, C. Fenouillet-Beranger, P. Batude, *et al.*, *IEEE International Electron Devices Meeting (IEDM)*, San Francisco, CA, 7.2.1-7.2.4 (2018).
33. J. Derakhshandeh, M. R. T. Mofrad, R. Ishihara, *et al.*, *J. Korean Phys. Soc.* **54**, 432-436 (2009).
34. R. Ishihara, P. C. van der Wilt, B. D. van Dijk, *et al.*, *Proc. ECS 202nd Meeting, Thin Film Transistor Technologies VI* (2002).
35. J. G. Lisoni, A. Arreghini, G. Congedo, *et al.*, *IEEE International Symposium on VLSI Technology*, Honolulu, HI, 1-2 (2014).
36. G. Congedo, A. Arreghini, L. Liu, *et al.*, 2014 *IEEE 6th International Memory Workshop (IMW)*, Taipei, 1-4 (2014).
37. K. Huet, C. Boniface, R. Negru, and J. Venturini, *AIP Conference Proceedings* **1496**, 135 (2012).
38. M. Belyansky, in *Handbook of Thin Film Deposition (Fourth Edition)*, K. Seshan, D. Schepis, Editors, pp. 231-268, William Andrew Publishing (2018).
39. S. K. Kim, K. M. Kim, D. S. Jeong, *et al.*, *J. Mat. Res.*, **28**, 313-325 (2013).
40. Tech Insights, Advanced CMOS Essentials report ACE-1701-805 (2017).

ECS Transactions, 89 (3) 155-164 (2019)
10.1149/08903.0155ecst ©The Electrochemical Society

Cluster-Preforming-Deposited Si-rich W Silicide:
A New Contact Material for Advanced CMOS

Naoya Okada, Noriyuki Uchida, Shinichi Ogawa, and Toshihiko Kanayama

National Institute of Advanced Industrial Science and Technology (AIST),
Tsukuba, 305-8568, Japan

Reduction of electrical resistance of middle-of-line (MOL) is now a primary requirement for advanced CMOS. In this paper, we demonstrate that the cluster-preforming-deposited (CPD) WSi_n ($n \leq 12$) film is a promising contact material for S/D in CMOS; insertion of the WSi_n film reduces the electron SBH to 0.32 eV at W/n-Si and to 0.43 eV at Co/n-Si junctions. This SBH reduction is due to the fact that the WSi_n film has a low work function of 4.0 eV and effectively passivates the Si surface without forming an appreciable density of interface states, thereby releasing the Fermi level pinning at the Si surface. These junctions have an excellent stability against annealing up to 700 °C for W/WSi_n/n-Si and to 500 °C for Co/WSi_n/n-Si. Furthermore, this film significantly prevents the Cu diffusion with a TDDB barrier height of 1.33 eV under a 5 MV/ cm stress.

Introduction

For decades, progress in Si CMOS transistors have been continued by dimensional shrinking, adopting new device structures, and introducing various novel materials. Many issues should be solved to keep this progress further. One of the key challenges is to reduce the electrical resistance of middle-of-line (MOL). The transistor performance is becoming highly dependent on the external series resistance and the contact resistance at source/drain (S/D) with recent shrinkage of Si CMOS. An obvious way to reduce the external series resistance is to replace the conventional W electrode by a more conductive metal, such as Co (1−3). To reduce the specific contact resistivity, low Schottky barrier height (SBH) at the metal/semiconductor interface is required. It is well known that for most metallic species, silicide/Si junctions always show a high electron SBH to n-Si owing to the Fermi-level pinning in the range between the mid gap and the valence band edge (4). This fact imposes a fundamental difficulty for obtaining low contact resistance and high performance in n-FET. Many researchers have already addressed this issue using various silicide and inserting a thin dielectric material to control the Fermi-level pinning (5); yet, an unconventional contact material is strongly required.

To meet this requirement, we are recently working on the Si-rich W silicide (WSi_n, n = 8−12) film as a contact material that does not form an appreciable interface states on the Si surface (6−12). The WSi_n film is composed of W-atom-encapsulated Si_n cage clusters. In the cluster, the W atom is connected with the surrounding Si atoms by strong covalent bonds (13, 14). When these clusters are aggregated into the film, a non-equilibrium phase results in an amorphous Si network by random connection of the clusters (15, 16). In the film, unlike the equilibrium WSi_2 phase, each W atom is separated from other W atoms by Si atoms and all the Si atoms are saturated with covalent bonding leaving few dangling bonds (Fig. 1). This unique film is expected to have an intermediate electronic state between metallic W and semiconducting Si, smoothly connecting Si with a metal electrode both chemically and electronically.

In this paper, we describe the deposition method and properties of the WSi_n film, and then demonstrate the SBH reduction effect by the WSi_n insertion. Additionally, we show another superior feature of the film: enough barrier properties against Cu diffusion.

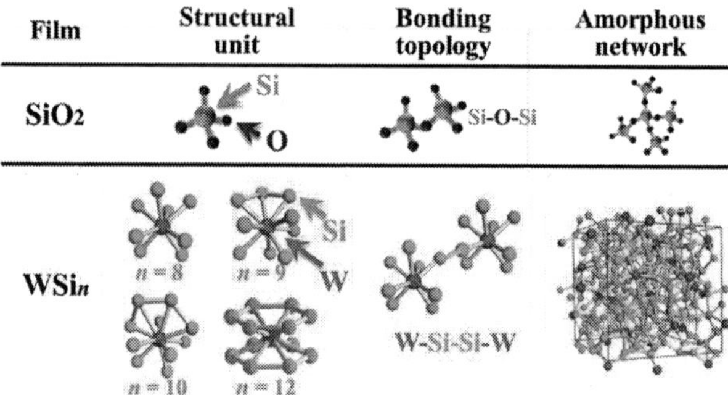

Figure 1. Schematics of structural unit, bonding topology, and amorphous network of the WSi_n film, in comparison with the conventional amorphous SiO_2 film. The structural units and amorphous network structure of WSi_n are calculated by density functional theory (13−16).

Cluster-Preforming Deposition (CPD) and Flm Properties of WSi_n

The WSi_n film was prepared by the newly developed cluster-preforming deposition (CPD) method (Fig. 2) (10). In a hot-wall thermal deposition system, hydrogenated W-atom-encapsulated WSi_nH_x clusters with well-controlled n values ($n \leq 12$) are preformed by reaction of WF_6 and SiH_4 in the gas phase. For the initial stage of the reaction ($n \leq 6$), the n values of clusters are determined by collision frequency of the WF_6 molecule with SiH_4, i.e., n depends on the SiH_4 pressure (10). In contrast, under conditions that promote sufficient collisions of clusters with SiH_4 (a partial pressure ratio of WF_6 to $SiH_4 < 1/10^3$), n is determined by the gas temperature controlled by the wall heater (10). We attribute this gas temperature dependence to the mechanism that the energy barrier for reaction between the clusters and SiH_4 increases with n (17). Actually, we observed in experiments using ion traps that the reaction probability of SiH_4 significantly deceases when the $WSi_nH_x^+$ cluster ions become larger than $n \geq 6$ and further slows as n increases (14, 22). The cluster is therefore a gas-phase product entirely different from the powder formation (17). The WSi_nH_x clusters are then deposited onto a substrate, where they are thermally dehydrogenated and coalesce to the WSi_n film with less hydrogen content and high atomic density without vacancies and voids (Fig. 3 (a)), as predicted by theoretical simulations (15). The residual fluorine concentration in the film is less than ~0.1 at. % because of the sufficient reduction reaction of WF_6 with SiH_4 in the gas phase (Fig. 3 (b)) (17). The formed films have an amorphous structure similar to the hydrogenated amorphous Si (10), but they are extremely stable against annealing up to 1000 °C (11). We also confirmed that the WSi$_n$ film exhibits an excellent step coverage over a hole array of 40-nm diameter with an aspect ratio of ~50; a completely conformal coverage is achieved down to the hole bottom (9).

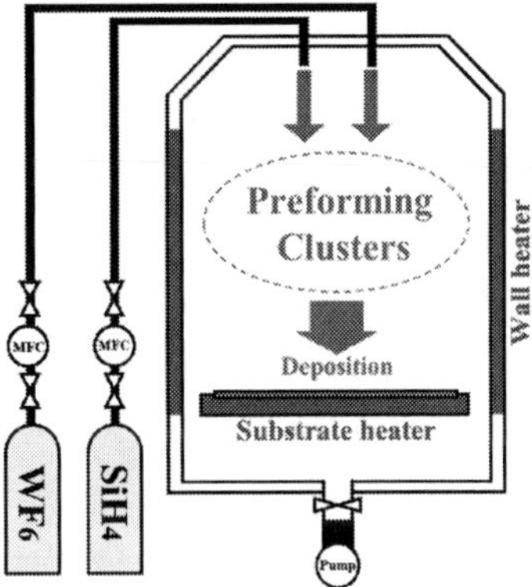

Figure 2. Schematic of the Custer-preforming deposition (CPD) system for the WSi$_n$ film. CPD forms the film in a non-equilibrium phase, composed of clusters with a well-defined composition and structure.

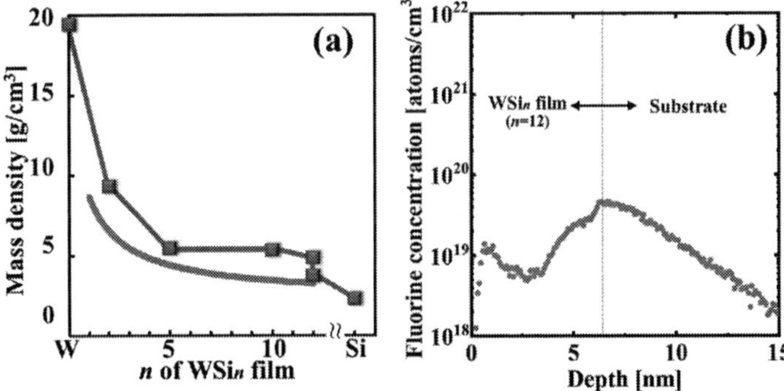

Figure 3. (a) Mass density of the resulting WSi$_n$ films, estimated by X-ray reflectivity (XRR) analysis, as a function of composition n. The blue curve is the mass density assuming the same atomic density as that of Si. The fact that all the data are above the curve indicates that the film has an atomic density higher than Si. The WSi$_n$ film is therefore densely packed with clusters without vacancies and voids. (b) Secondary ion mass spectrometry (SIMS) depth profile of the residual fluorine in the WSi$_n$ film with n =12. The residual fluorine in the film was very few of < ~0.1 at. % because of the sufficient reduction reaction with SiH$_4$ in the gas phase. This value is much lower than that in a conventional CVD film of > ~1 at. % (18).

Junction Properties of Schottky Diodes with WSi_n Insertion

To simulate S/D contacts in n-FET, we evaluated the effect of WSi_n film insertion into Schottky diodes with a high resistivity Si substrate: $W/WSi_n/n\text{-Si}$ (~2 Ω·cm). Here, the WSi_n films of 5–10 nm thickness and the W electrodes were successively formed in the same CPD system using the same source gases. Figure 4 shows the electron and hole SBHs estimated by capacitance-voltage ($C\text{-}V$) characteristics at room temperature (RT) as a function of n (9). For the direct junction without WSi_n, the electron SBH was 0.67 eV due to the Fermi-level pinning near the charge neutrality level of Si, which is located at an intermediate level between the mid-gap and the valence band edge of Si. By inserting the WSi_n with $n = 6$, the electron SBH was reduced to 0.58 eV and further lowered to 0.45 eV when $n = 12.3$. At the same time, the hole SBH increased; the sum of electron and hole SBHs were kept at ~1.1 eV, equal to the Si band gap, verifying the same situations are kept for both n and p junctions. This indicates that the Fermi-level pinning is released without forming an appreciable density of interface states. This passivation by WSi_n becomes more effective with the larger n values. In addition, the annealing treatments (in N_2 gas at 700 °C for 30 min) provided a more reduction of electron SBH to 0.32 eV when $n = 12.3$. This is also probably due to the reduction of interface states on the Si surface.

The WSi_n insertion may do harm to p-FET junctions because the hole SBH increases as shown in Fig. 4. However, we found that this issue is solved when a Ge layer is used in S/D for p-FETs (9). The epitaxial Ge or SiGe is widely used as a S/D material to lower the contact resistance and exert compressive strain to the channel. For the $W/WSi_n/Ge/p\text{-}$Si junction, the hole SBH was reduced with annealing temperature, approaching the charge neutrality level close to the valence band edge of Ge; e.g. the hole SBH = 0.51 eV with $n = 12$ after annealing at 600 °C. This is because Ge atoms at the interface prefer to mix with WSi_n and the annealing leads to an increase of interface state density on Ge, resulting in strong Fermi-level pinning at the valence band edge. Thus, the depinning effect of the WSi_n film is lost in the annealed Ge junction, while it is preserved in the Si junction. This property is particularly advantageous in that the insertion of common WSi_{12} film reduces both the electron and hole SBHs simultaneously.

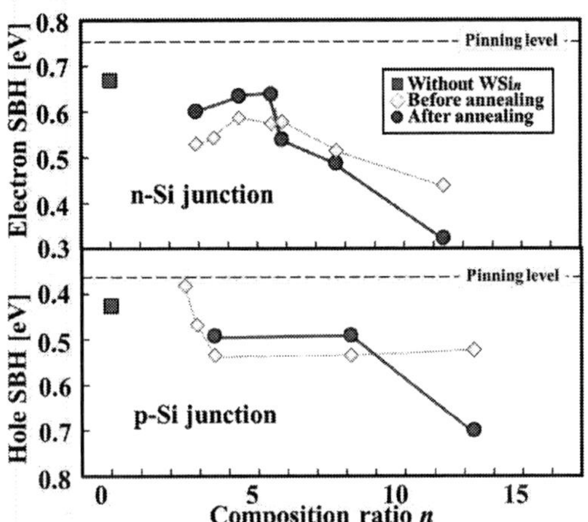

Figure 4. Electron Schottky barrier height (SBH) of the W/WSi$_n$/n-Si and hole SBH of the W/WSi$_n$/p-Si estimated from capacitance-voltage (*C-V*) characteristics as a function of composition n before and after the annealing treatment at 700 °C.

We next focused on the Co electrode, which is now attracting strong attention as an alternative to W or Cu for local connections and interconnections (1–3), while it has a high diffusivity in Si and a low reaction temperature easy to form silicide with Si. The Co/Si junction therefore needs a barrier layer against Co diffusion (19). To evaluate both the SBH reduction effect and the barrier properties against Co diffusion, we investigated the thermal stability of the Si Schottky diode with a high resistivity n-type Si substrate (~2 Ω·cm), inserting the WSi$_n$ film of 5–20 nm thickness: Co/WSi$_n$/n-Si. Here, the Co electrode of 50 nm thickness was sputtered at RT onto WSi$_n$. The Co/n-Si direct junction showed a high electron SBH of 0.64 eV owing to the Fermi-level pining at the Si surface (Fig. 5). For the Co/WSi$_n$/n-Si junction, the electron SBH gradually lowered with increasing n of WSi$_n$ and reached to 0.43 eV when n = 8.3. Although a slight increase in SBH was observed for the junctions with $n \geq 6$ after the annealing at 500 °C, this SBH reduction was kept even after the annealing at 500 °C. Note that the annealed Co junctions exhibit SBH values similar to the W/WSi$_n$ junctions. The apparent increase in SBH may be due to the fact that the as-deposited Co junctions show a large scattering in SBH values probably because the Co electrodes were sputter deposited in a different chamber after exposing the samples in air, while the W electrodes were successively formed in the same CPD system. The above result indicates that the Co silicidation with the Si substrate was prevented by the WSi$_n$ insertion at least up to 500 °C and hence the WSi$_n$ worked as a retarding layer against Co diffusion. This barrier effect is because in the WSi$_n$ film, all the Si atoms are strongly bonded to the W atoms forming W-atom-encapsulated clusters, retarding the interdiffusion of Si atoms of the WSi$_n$ and Co atoms.

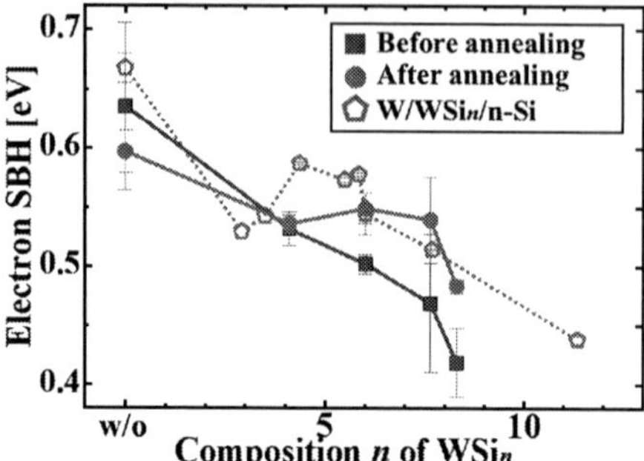

Figure 5. Electron SBH of the Co/WSi$_n$/n-Si junctions estimated from C-V characteristics at RT as a function of composition ratio n of the WSi$_n$ film before and after annealing at 500 °C, in comparison with W/WSi$_n$/n-Si junctions before annealing.

The above junction properties indicate that the WSi$_n$ film has a low work function. To verify this fact, we investigated the effective work function (eWF) of the WSi$_n$ film by C–V measurements of metal-oxide-semiconductor (MOS) capacitors of W/WSi$_n$/SiO$_2$/n-Si layered structures as a function of composition n in the range of 4.1−12.3. The eWF was obtained from the sum of the flat band voltage V_{fb}, the electron affinity of 4.05 eV in Si, and the difference between the conduction band edge E_c and the Fermi level of 0.25 eV in the n-Si substrate. The eWF of the W/SiO$_2$/n-Si was estimated to be 4.4−4.8 eV, consistent with the work function of the W electrode. On the other hand, the eWF of WSi$_n$ decreased with n, and reached to 4.0 eV when $n > 6$. This eWF is close to the E_c, indicating that the Fermi level of WSi$_n$ is located at this position. These eWFs remained in the same position after annealing up to 600 °C (Fig. 6). Our previous work showed that the WSi$_n$ film remains in the same amorphous structure on SiO$_2$ up to 1000 °C (11). Therefore, the Fermi level of WSi$_n$ will be stable unless the film structure changes. In general, a metallic material with a low work function is unstable (20). Actually, for most of metals with eWF values lower than the mid gap of Si, annealing tends to increase the eWF towards the charge neutral level (Fig. 6) (21). The eWF behaviors are significantly different from those of SBH; eWF reaches E_c when $n > 6$ but SBH gradually changes and does not reach the band edge. This indicates that the Fermi-level pinning is still working slightly at the Si interface in the Schottky diodes. In fact, this interface pinning effect was verified by the fact that the annealing treatment is effective to SBH (Fig. 4) but is not to eWF (Fig. 6).

To elucidate why the WSi$_n$ film has a low eWF, we simulated the distribution of electrical charge on each atom in a structure assembled from ten WSi$_{12}$ clusters using density functional theory calculations. As shown in Fig. 7, all the Si atoms form cage structures surrounding the W atoms, preserving the property of the isolated cluster. In this structure, electrons are transferred from p orbitals of the Si atoms to the W atoms to fill the d orbitals. This mechanism raises the Fermi level and results in a low eWF.

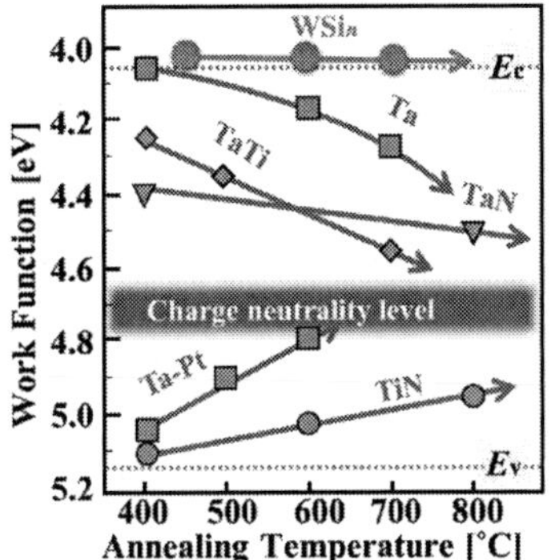

Figure 6. eWF of WSi$_n$ (n = 8.3) in the W/WSi$_n$/SiO$_2$/n-Si capacitors before and after annealing in the temperature range of 400−700 °C. The eWFs of various Metal/SiO$_2$/n-Si capacitors are cited from ref. (21).

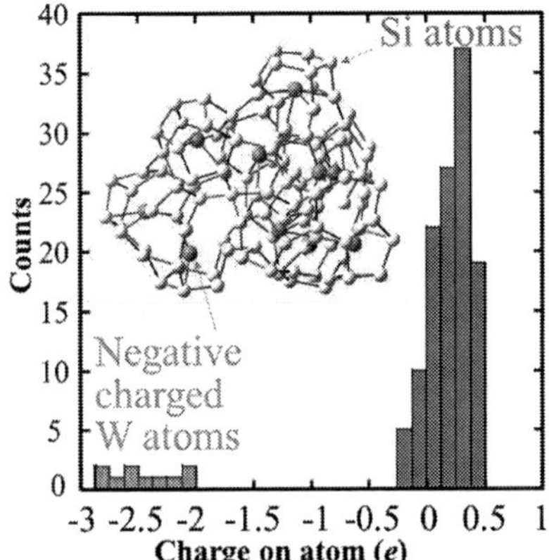

Figure 7. Distribution of electrical charge on each atom calculated in the structure assembled from ten WSi$_{12}$ clusters. All the W atoms are negatively charged by electrons transferred from the Si atoms.

Diffusion Barrier Properties for Copper

Although Cu is widely used as an interconnect metal in LSIs, Cu is well known to have a very high diffusivity in Si and SiO$_2$. Thus, we investigated the barrier properties of the WSi$_n$ film against Cu diffusion using the two test structures of Cu electrodes: MOS capacitors (thickness t of the thermal oxide (SiO$_2$) = 20 nm) and p$^+$-n diodes (doping concentration of p$^+$-Si = ~10^{20} cm^{-3} to a depth of 100 nm, that of underlying n-Si substrate = 2×10^{14} cm^{-3}) with and without the WSi$_n$ (n = 12, t = 5 nm). Ta (t = 100 nm)/Cu (t = 200 nm) electrodes with diameters of 100−500 μm were sputtered at RT onto WSi$_n$, SiO$_2$, or Si. The MOS breakdown time was evaluated under a bias-temperature stress (BTS) of 5 MV/cm at 200 °C, 225 °C, and 250 °C in a vacuum of ~5×10^{-5} Pa. The Cu/SiO$_2$/Si capacitor without the WSi$_n$ film failed at a much earlier stress time than the one with WSi$_n$ (Fig. 8 (a)). The Arrhenius plot of break-down time shows a time-dependent-dielectric-breakdown (TDDB) barrier height of 1.33 eV under the stress condition of 5 MV/cm (Fig. 8 (b)) (12).

For the p$^+$-n diodes, the ideality factor η was calculated from the *I-V* characteristics with repeating 30-min isochronal annealing in a N$_2$ atmosphere at 400–600 °C. The p$^+$-n diodes with the WSi$_n$ film remained in η = ~1, an ideal value even after the annealing, while the η value of that without WSi$_n$ widely scattered increasingly with the annealing temperature (Fig. 9), indicating the increase of recombination current resulting from the Cu diffusion. Thus, these results verify that the WSi$_n$ film has sufficient barrier properties for direct Cu contact. This excellent stability was very easy to observe that the surface color changed by the annealing only on the electrodes without WSi$_n$.

This barrier mechanism is attributed to the low work function of WSi$_n$, mentioned above (Fig. 6). The low work function leads to the prevention of charge transfer from Cu (work function: ~4.6 eV) to WSi$_n$ (work function: ~4.0 eV). In contrast, it is well known that Cu is easily ionized in Si, and resulting Cu$^+$ ions have very high diffusivity.

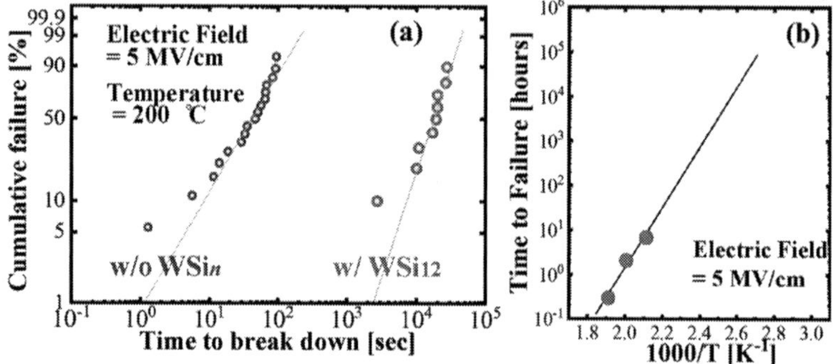

Figure 8. (a) Weibull plot of MOS capacitors under 5 MV/cm at 200 °C and (b) Arrhenius plot of the time to failure under 5 MV/cm stress for Cu/WSi$_{12}$/SiO$_2$/Si.

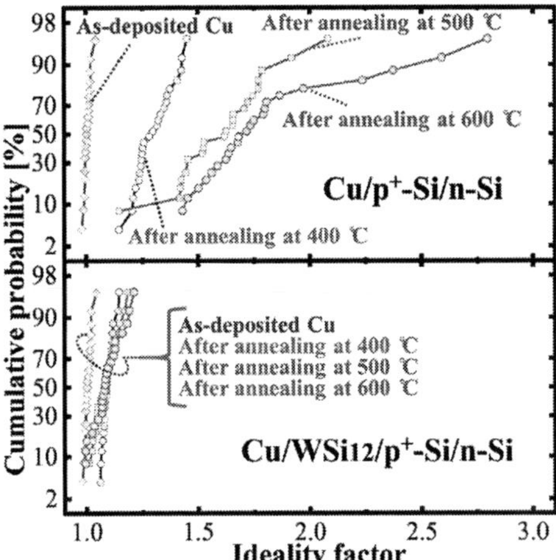

Figure 9. Distributions of the ideality factor in Cu/p$^+$-Si/p-Si and Cu/WSi$_n$/p$^+$-Si/p-Si diodes before and after annealing in the temperature range of 400–600 ℃.

Conclusions

We demonstrated that the insertion of the thin amorphous WSi$_n$ film with $n \leq 12$ reduces the electron SBH, and furthermore this film exhibits excellent barrier properties against Cu diffusions. The major conclusions are as follows. These features ensure that the WSi$_n$ film is well suited to the contact material in the most advanced CMOS.

1) The amorphous WSi$_n$ film is prepared in a hot-wall thermal reactor using WF$_6$ and SiH$_4$ gas sources: *i.e.*, the cluster-preforming deposition (CPD) method. The hydrogenated WSi$_n$H$_x$ clusters, synthesized in the gas phase, land on a heated substrate, where they dehydrogenate and coalesce into a dense film with less hydrogen and fluorine content. The WSi$_n$ film is formed with an excellent contact hole coverage.

2) The WSi$_n$ film has an excellent thermal stability that the same amorphous structure remained up to 1000 °C in spite of the Si-rich composition.

3) The WSi$_n$ film has a low eWF of 4.0 eV and its insertion reduces the electron SBH in the Si Schottky diode: 0.32 eV for W/WSi$_n$/n-Si junction and 0.43 eV for Co/WSi$_n$/n-Si junction. These junctions have the excellent stability against annealing up to 700 ˚C for W/WSi$_n$/n-Si and to 500 ˚C for Co/WSi$_n$/n-Si junctions.

4) The WSi$_n$ film exhibits excellent diffusion barrier properties for Cu: a TDDB barrier height of 1.33 eV under a 5 MV/cm stress for Cu MOS capacitors and a high stability of Cu on Si diodes against annealing up to 600 °C.

Acknowledgments

This research was supported by JST PRESTO No. JPMJPR1326 and JSPS KAKENHI No. JP18K13794.

References

1. R. Hung, et al, *IEEE IITC*, p. 30, (2018).
2. A. Yeoh, et al., *IEEE IITC*, p. 144, (2018).
3. C. Auth, et al., *IEDM Tech. Dig.*, p. 673, (2017).
4. T. Nishimura, K. Kita, and A. Toriumi, *Appl. Phys. Lett.* **91**, 123123 (2007).
5. A. Toriumi, and T. Nishimura., *Jpn. J. Appl. Phys.* **57**, 010101 (2018).
6. N. Okada, N. Uchida, and T. Kanayama, *J. Appl. Phys.* **117**, 095302, (2015).
7. N. Okada, N. Uchida, and T. Kanayama, *Appl. Phys. Lett.* **104**, 062105 (2014).
8. N. Okada, N. Uchida, and T. Kanayama, *Appl. Phys. Lett.* **101**, 212103 (2012).
9. N. Okada, N. Uchida, S. Ogawa, and T. Kanayama, *IEDM Tech. Dig.*, p. 553, (2017).
10. N. Okada, N. Uchida, and T. Kanayama, *J. Chem. Phys.* **144**, 084703 (2016).
11. N. Okada, N. Uchida, and T. Kanayama, *J. Appl. Phys.* **121**, 225308, (2017).
12. N. Okada, N. Uchida, S. Ogawa, and T. Kanayama, *IEEE IITC*, p. 132, (2018).
13. N. Uchida, T. Miyazaki, and T. Kanayama, *Phys. Rev. B* **74**, 205427 (2006).
14. H. Hiura, T. Miyazaki, and T. Kanayama, *Phys. Rev. Lett.* **86**, 9 (2001).
15. T. Miyazaki, N. Uchida, and T. Kanayama, *Phys. Stat. Sol. C* **7**, 636 (2010)
16. N. Uchida, T. Miyazaki, Y. Matsushita, K. Sameshima, and T. Kanayama, *Thin Solid Films* **519**, 8459 (2011).
17. N. Okada, N. Uchida, S. Ogawa, and T. Kanayama, *Jpn. J. Appl. Phys.* **58**, SBBA09, (2019).
18. T. T. Kodas and M. J. Hampden-Smith, *The Chemistry of Metal CVD* (Wiley-VCH, 1994), p. 105.
19. M. Hosseini, D. Ando, Y. Sutou, and J. Koike, *Microelectron. Eng.* **189**, 78, (2018).
20. R. Kuroda, et al., *IEDM Tech. Dig.*, p. 580, (2010).
21. H. Y. Yu, et al., *Ext. Abstr. Solid State Devices and Materials*, p. 712, (2004).
22. A. Negishi, N. Kariya, K. Sugawara, I. Arai, H. Hiura, and T. Kanayama, *Chem. Phys. Lett.* **388**, 463 (2004).

ECS Transactions, 89 (3) 165-166 (2019)
10.1149/08903.0165ecst ©The Electrochemical Society

Author Index